Tracing EMFs
in Building Wiring
And Grounding

*A practical guide for reducing magnetic fields
due to wiring errors as well as
current grounding practices*

Third Edition, Revised

Karl Riley

Dedicated to

all those who have crawled through spider webs under houses or
through loose insulation in the attic in order to fix a mis-wired circuit,

and to all those concerned homeowners and tenants who were unable to get
good serious factual responses from the various authorities they consulted.

ISBN: 1-4699-0201-X
ISBN-13: 9781469902012

Published by **Karl Riley,** PO Box 441, West Tisbury, MA 02575;
kriley3@earthlink.net

TABLE OF CONTENTS

INTRODUCTION

What this guide is about

This guide is for electricians and electrical inspectors, those starting a new profession in EMF consulting, and anyone involved with the process of building construction, such as architects, contractors and home-owner electricians. Utility personnel involved in EMF measurements can profit from this information.

The guide will explain a few basic principles of AC magnetic field production in relation to current cancelation and conductor spacing.

Next you will be given an overview of how elevated magnetic fields are created by certain building and wiring practices, many of which are in violation of the National Electrical Code (NEC).

There is a chapter devoted to clearing up misconceptions about grounding. Following this is a discussion of detection and measurement of AC magnetic fields in buildings, followed by a chapter on locating the strong field sources or circuits.

Next you will be shown how to diagnose which circuits are involved by taking measurements at the electrical panel boxes or out in the circuits.

The basic purpose of this book is to help you successfully locate the wiring fault or grounding connection that is causing the strong fields.

Next a number of case histories are described where wiring and grounding methods caused high magnetic fields until diagnosed and corrected.

There is an appendix on instrumentation and a bibliography for further reading on this subject. See Appendix A for info on my dvd which illustrates the main themes of this book.

The terminology in this guide is a compromise between electricians' terminology and wording that other professionals may understand. A glossary clarifies definitions and provides more information. The reader is encouraged to make use of it.

PREFACE TO SECOND EDITION, REVISED

When the first edition of this book was published in 1995 there was very little awareness on the part of electricians of the specific wiring errors which cause high magnetic fields. Ten years later there has been a change. Some of this change is due to the spread of this book; some is due to the magnetic field survey of 90 California schools funded by the California Health Services, in which it was found that wiring error was the most frequent source of elevated magnetic fields in schools; and some of the change is due to Mike Holt's initiative in inviting me to become a moderator on the NEC® Code Forum on his website, www.mikeholt.com . This is the most used website for electricians discussing Code matters in the world, and the effect of various Code violations on magnetic field production is now well understood and intelligently discussed.

One of the most frequent topics discussed is Grounding and Bonding, and the Forum deals with these questions daily. When the topic involves increased magnetic fields I get involved, but increasingly I find that others speak up and give sound information. Whereas a few years ago I would have felt obliged to bring up this aspect of grounding, now I find that someone else, whether a moderator or poster, has answered the inquiry as well or better than I would have. That is very pleasing to see.
At the same time, we have a long way to go. I quote an electrical instructor who asked me to withhold his name:

> Mr. Riley
> The biggest problem I have encountered with EMF is the attitude of electrical contractors. I have taught the electrical apprenticeship program the last three years. Every year I have to go through the proper way to wire three-way switches, ground and neutral connections, etc. I keep getting the same answers from the apprentices,"this is the way the boss wants it". We may be teaching the wrong ones. The class should be for the ill-informed master electricians and business owners. The problem also encompasses inspection and community leaders who are more interested in being good ole boys and votes than electrical safety. I will admit though things are improving. Thanks for your work in this area.

The book has also gone abroad, and I have had positive feedback from England, Australia, Canada and Russia, to name a few. I recently received an order for the book from a police Constable on Mauritius (off the East African coast) for his course in Public Health.

I really appreciated the following email from a Victor Petoukhov from Russia, which began:

> Dear Karl,
> at last (a lot of job!!) I have read your wonderful book!
> I am ready to sign under it's every word.

Yes, it was a lot of job, and I am happy to send another edition on its way.

PREFACE TO THE THIRD EDITION, revised 2012

Few changes have been necessary when updating to 2012. Building wiring when done following the National Electrical Code continues to produce low magnetic fields, with the exception of some grounding practices. However, some electricians and home owners also continue to make errors or use shortcuts which result in high magnetic fields, which research shows may be hazardous to our health. Many more books are now available on the internet dealing with health research as well as technical aspects of magnetic field production. More and sometimes cheaper gaussmeters are available. The battle continues between information and mis-information. May this book play its part in showing how to make simple and logical corrections to eliminate high magnetic fields.

Chapter 1

THE PROBLEM

Elevated magnetic fields cause disturbances in sensitive electrical equipment, induce currents in conductors with accompanying heat, and are associated with some disease conditions. If you have purchased this book you may already be aware of some of the health concerns which have focused attention on AC magnetic fields and their causes.

The research on the association between certain diseases and magnetic fields is on going and is considered a controversial area. Some of the controversy is from within the scientific community, but as with any health concern which threatens to shake up some industries, much of the controversy is bound up with corporate interests and governmental budgetary concerns. On the other side are citizen activists who want remedial action during this period of scientific uncertainty. The National Institute of Environmental Health Sciences (NIEHS) and the U.S. Department of Energy (DOE) tried in 1995 to present a balanced view of the health situation in a recent booklet: *Electric and Magnetic Fields Associated with the Use of Electric Power* (1995). To quote from the Introduction:

> Recent epidemiological studies have suggested that a link may exist between exposure to power frequency electric and magnetic fields (EMFs) and certain types of cancer, primarily leukemia and brain cancer. Laboratory research is under way to determine whether such an association is biologically possible.

> Several epidemiological studies have looked for EMF effects on pregnancy outcomes, the nervous system, and general health. Various EMF sources have been studied for possible association with miscarriage risk: power lines and substations, electric blankets and heated water beds, electric cable ceiling heat, and computer monitors or video display terminals (VDTs).........p. 20

> Several studies looked at the overall health of high voltage electrical workers, and a few looked at the incidence of suicide or depression in people living near transmission lines. Some studies have also investigated the possibility that certain sensitive individuals may experience allergic-type reactions to EMFs.

> Research reported in 1994 has suggested a possible link between occupational EMF exposure and increased incidence of Alzheimer's disease. . . .p. 20

In the years since that booklet was issued, literally hundreds of studies have been conducted on biological effects of power frequency magnetic and electric fields.

So many epidemiological studies hjave been conducted that they can be combined (with adjustments) in order to increase the numbers (cases and controls). Increasing the population numbers allows the statistical correlations to be more significant and trustworthy.

One such mega-study was conducted by the International Agency for Research on Cancer (IARC), an agency of WHO (World Health Organization), published in 2001. The conclusion of this conservative body was that a doubling of childhood leukemia is associated with an average AC power frequency magnetic field level of 4 milligauss (mG). [This equals 0.4 μT, or zero point four micro Tesla as used in Europe]. Thus a designation of "possible carcinogen" (2-b) was given to this field strength and above.

As a frame of reference I have measured many bedrooms and school classrooms with fields of 4mG or higher due to mistakes in wiring connections.

It is an interesting coincidence that this is also the field level where some computer screens will begin to jitter.

For those interested in reading the various viewpoints about the health issues consult the Bibliography.

As electricians you will be called in with increasing frequency to correct errors which have shown up on someone's gaussmeter, to advise about grounding problems and replace legal (grandfathered) but outdated knob-and-tube wiring. What may have started out as a customer's health concern usually shows itself as a problem requiring correction from the point of view of correct wiring. In other words when a Code violation is involved, fire and shock hazards also exist.

As an EMF consultant I am called in to trouble shoot wiring errors which cause interference with electronic equipment almost as often as when the main concern is health. Sometimes the two are present together, though the health concern may be kept in the background due to prevailing attitudes.

Chapter 2

HOW THE WIRING PROBLEMS SURFACED

When I first started conducting magnetic field surveys in 1989 the concern was centered on power lines, but when I walked into buildings that were well away from power lines I often found magnetic field levels which were as high or higher than power line fields.

In a friend's house on Marthas Vineyard, Mass. I found a field of 32 mG (milligauss) running along the floor through the living room. (Nationwide residential median is 0.5 mG according to one study by EPRI – Electric Power Research Institute). In one elementary school I found a similar field running through a 5th grade classroom, through the hall and up a wall. In a high school there were high fields in the hall coming from the sprinkler pipes above and a source under the floor below. Many classrooms were affected.

I worked with local electricians, and we traced the sources. My friend's 32 mG was due to the grounding of a neutral bus in a subpanel to a disconnected water pipe from a previous water supply coming from an adjacent building. Code violation. (Neutral buses in sub-panels may not be grounded).

The field coming from the floor in the 5th grade classroom was due to the bonding of neutral to ground at a subpanel in the hall. Code violation.

The currents on the sprinkler pipes in the high school causing high fields in the hall and some classrooms were due to neutral/ground bonds in the subpanels, an intentional neutral/ground shunt on a receptacle and a connecting electrical path through the metal wall lathe to the sprinkler pipes. More Code violations.

As I continued my surveys, other electrical connections turned up that were causing elevated fields. It was necessary to become very familiar with the National Electrical Code, particularly Article 250, and to study authoritative works on grounding and bonding such as the IAEI's Soares Book on Grounding. (IAEI = International Association of Electrical Inspectors).

Years later I found a treasure trove of information in the Code Forum on Mike Holt's website, www.MikeHolt.com After the first edition of this book was published in 1995, Mike Holt contacted me and gave me a thumbs up on the book. He had been emphasizing some of the same points I was focussed on, particularly in relation to grounding. Eventually Mike invited me to join the Moderators of the NEC Code Forum and share my particular angle on wiring and EMF production. I have been doing this for a few years now and have seen some good changes of attitude and engaged in many vigorous and productive discussions with experienced electricians and engineers. This website is an extraordinary learning vehicle and I recommend it highly to electricians. When a question is posted it is answered and discussed the same day. It is not for DIYers but all can read and follow the discussions. Moderators serve on a voluntary basis. They are a great bunch of guys, each with his particular area of expertese.

To return to my story, while working with electricians I learned the reasons why some of the errors we were finding had been made. Whether the electrician had been licensed or unlicensed did not seem to have much bearing on whether he had decided to cut corners, but may have had something to do with errors made out of ignorance of the Code and of the electrical principals on which Code is based.

Figure 1 tabulates the sources of magnetic fields of over 3 mG in substantial living areas of the first 150 buildings I surveyed, both on the East and West coasts. Subsequent surveys have shown this source allocation to remain typical. This sample is actually biased towards power line fields, since clients would often call me in because of a nearby power line or transformer.
The second graph in **Figure 1** is a breakdown of the wiring/grounding category which shows that approximately 66% of these fields were due to wiring errors.

The graph at **Figure 1b** is a breakdown of the power line category, showing that primary distribution lines are by far the most common source of elevated magnetic fields as compared with transmission lines and distribution secondary. Since magnetic field strength is a function of current load as well as proximity (and line configuration) this is not surprising, as not that many buildings are located near transmission lines as compared with the network of primary distribution lines that carry current to all sections of towns and cities.

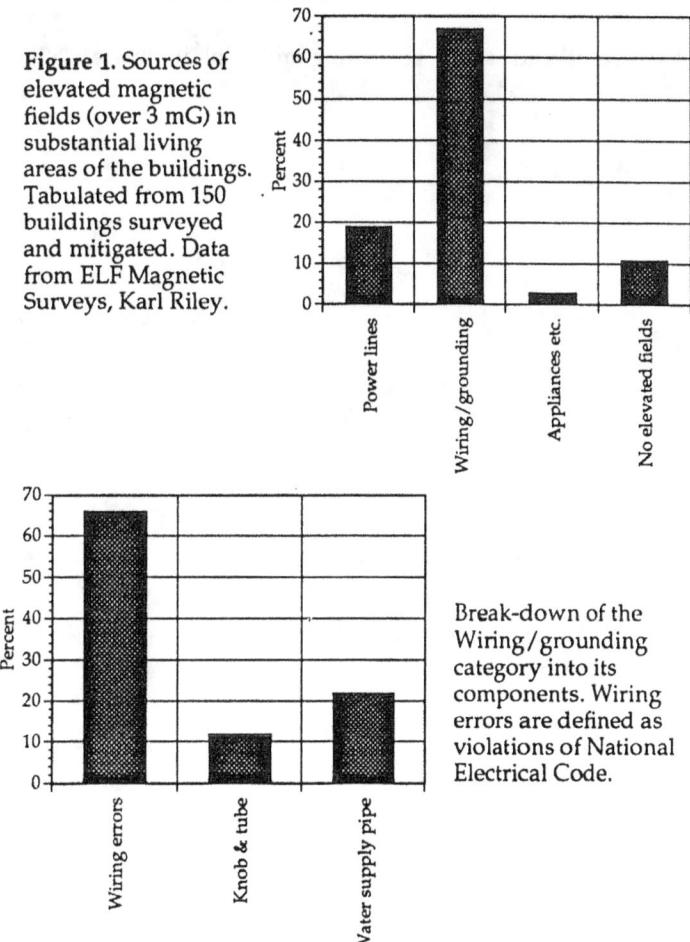

Figure 1. Sources of elevated magnetic fields (over 3 mG) in substantial living areas of the buildings. Tabulated from 150 buildings surveyed and mitigated. Data from ELF Magnetic Surveys, Karl Riley.

Break-down of the Wiring/grounding category into its components. Wiring errors are defined as violations of National Electrical Code.

Break-down of Power line source category into its components.

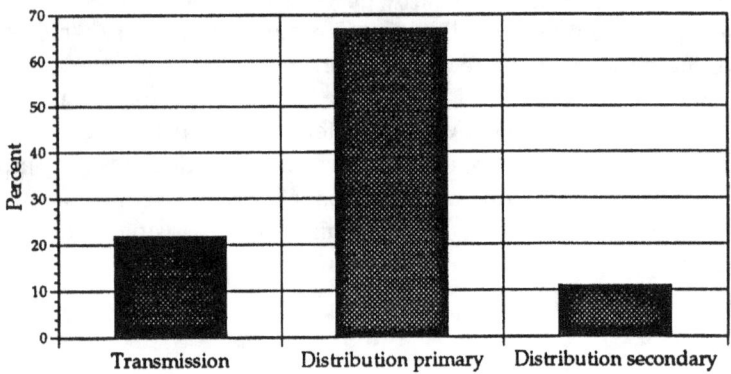

Figure 1b. When power lines were the main source of the magnetic fields.

The next chapter will start at the beginning and explain how an "elevated" magnetic field is created.

Chapter 3

WHAT CAUSES ELEVATED MAGNETIC FIELD LEVELS?

If a person with a gaussmeter [measures magnetic fields in milligauss (mG)] walks through a one or two story building with normal electrical loads, the reading on the meter's display will usually be from 0.0 mG (this means less then one tenth of a milligauss) to about 0.3 mG if there are fluorescent lights on.

If the fields are generally higher the source or sources may be outside the building. This is easy to check. If the source/s are within the building we usually find either incorrect wiring connections or neutral currents on water pipes within the building due to certain grounding requirements.

In other words, in a correctly wired building not too close to a power line I normally see only a few tenths of a milligauss in the major spaces, and this is usually from the fluorescents overhead. Now that higher frequency electronic ballasts are being used more frequently, we may see nothing on our meter from the fluorescents.

If the building is miswired in certain ways it is common to measure fields of 2-30 mG in some rooms and much higher near some walls, floors or ceilings.

If one measures near the surface of computers, fluorescent lights or any motor or transformer the field level may go up into the hundreds or even thousands of milligauss. But it weakens rapidly as we will see.

Let's take a basic look at what causes AC magnetic fields. We are not concerned with the steady magnetic fields such as the earth's field or the field from a DC current. These are not associated with the biological effects that the laboratory studies show from alternating fields, because DC does not induce currents and AC does.

How are AC magnetic fields created?

When an electric current flows a **magnetic** field always accompanies it. You could say it is the magnetic component of a unitary event.

An **electric** field is present around any source carrying voltage whether there is current flowing or not, but a **magnetic** field is created only when current flows. Electric fields are easily shielded by building walls and any grounded object such as a tree. Magnetic fields are unaffected by most building materials and so walls and floors make no difference. In steel framed buildings there are some concentration and attenuation effects. Since AC in the US is 60 cycles per second, the current changes direction 120 times per second, and the magnetic field spreads out and collapses at the same rate.

When a magnetic field moves past a conductor it induces an electrical current in the conductor. As it collapses it induces an opposite current in the conductor. AC induces AC.

So the electric current creates a magnetic field, and a moving magnetic field creates an electric current. Move a magnet past a wire at right angles to the wire and you cause a momentary current to flow.

Move a magnetic field continuously past a lot of wires (a coil) and you have an electric generator. All the AC electricity we use is created by moving magnetic fields. It is transferred and its voltage stepped up or down by magnetic fields in transformers. Motors turn because of the magnetic fields generated. I mention all these elementary facts because I have listened to a utility engineer and EMF spokesman tell a public meeting that "magnetic fields contain no energy- they're just fields". Indeed! There does seem to be a degree of misinformation of a very basic nature floating around in parts of the electrical community.

Magnetic field proportional to current

The magnetic field's strength (technically "flux density") coming from a single line source is proportional to the current. It weakens with distance, but only directly with distance (twice the distance, one half the field strength). There is a simplified formula which you can use to predict the field if you know the amperage, or vice versa: Twice the current in amperes divided by the distance in meters = the magnetic field's strength in milligauss.

$2i \div r = B$ (Where "i" is current in Amps, "r" is distance in meters, and "B" is magnetic field strength in milligauss - mG). So a 16 amp current in a single conductor produces a magnetic field of 32 milligauss a meter away. $2 \times 16 / 1 = 32$ mG.

This formula can be used on a conductor like a current carrying water pipe, but electrical conductors run in pairs or groups of three, four or more. What happens in a cable with a hot and neutral each carrying 10 amps? Since the currents in each conductor are running in opposite directions, 180° out of phase, the magnetic fields cancel each other- not completely since there is a small space between conductors but well enough so that you will not even measure one or two tenths of a milligauss 1 foot away from the cable.

Terminology time out:
Notice that I said "hot and neutral". Actually the return conductor in this case is not technically a neutral, but a "grounded conductor" or "grounded circuit conductor". We have been used to calling the white wires "neutrals", but there is a push now to be more technically correct. Neutral actually applies to a return conductor in multi-phase circuits, or circuits with two hot legs, such as the usual service conductors feeding a building. It is correct for branch circuits usually called "3-wire", or when used with a three-phase circuit. It carries only the unbalanced vector sum current of the phases it runs with.

So what do we call the grounded conductor? Too big a mouthfull for constnt use. So I will use "GC" for grounded conductor in this book, and hope it does not cause any confusion. Remember that the bare or green wire is called the "equipment grounding conductor" and is usually referred to as the EGC for short. So the usual Romex cable contains a black hot, a white GC and a bare EGC.

Let's look at knob-and-tube wiring where the hot and GC are usually on adjacent studs, joists, etc., or even further apart. Now the distance between conductors is so great that they almost act as single conductors as far as magnetic field production goes. If they spread out and then come together a loop is created. The field is high and very uniform within the enclosed space. It can be calculated by the simple formula, $B = 4i \div d/2$, where B = mG, i = amps and d = distance between conductors in meters. In other words, double the field strength compared with a single net current source.

Figure 2 shows graphically the difference in the magnetic field produced by a 10.4 amp current flowing in paired conductors compared with 10.4 amps flowing in a single conductor such as we see in knob-and-tube wiring. Note that the field from the paired conductors is almost non-existent directly above the conductors, but is measurable to the side.

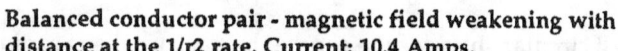

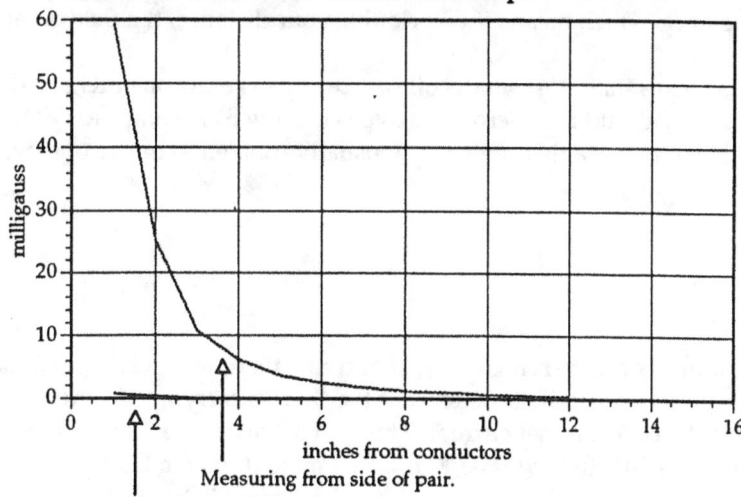

Measuring from side of pair.

Measuring directly above pair. The usual orientation is somewhere between these two curves.

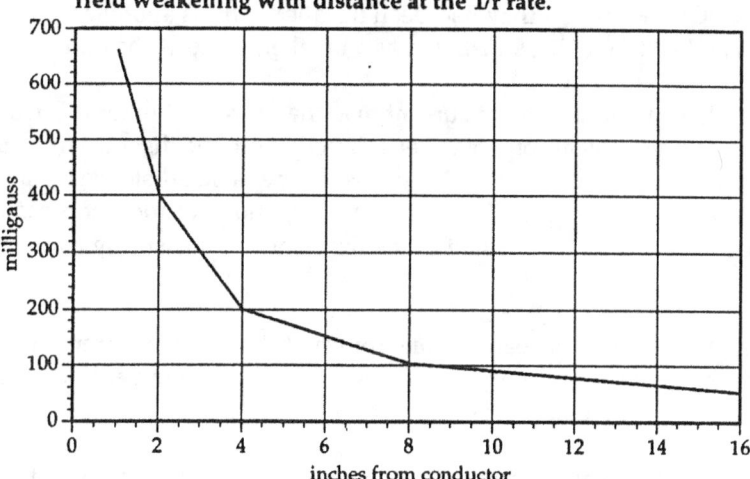

Figure 2. Comparison of net current of 10.4 A with a balanced pair at 10.4 A.

A coil multiplies the field

Since a current traveling through a wire creates a magnetic field, this effect can be multiplied by increasing the number of wires. You can also focus the field by using circular wires. This is called a coil, which produces a magnetic field in relation to the number of turns, the diameter of the coil, and the amount of current flowing through it.

Since motors contain coils and transformers contain coils and most appliances contain one or both, it is obvious why they can create very high magnetic fields. I have measured 31,000 mG (31 gauss) at the casing of an electric pencil sharpener in a classroom. Since most gaussmeters cannot measure fields this high, many technicians are not even aware such fields exist in public places. I was using an **MSI 95** single axis gaussmeter (see **Instrumentation** in Appendix).

Because of the small circular shape of the source the field weakens rapidly - basically with the cube of the distance, which means if you double your distance, the field weakens to one eighth.

In practice this means that most small coil sources drop below one mG at a distance of 18 inches; for some it may be 24" and for microwave ovens usually 5'. A computer VDT coil around the neck of the tube produces a field which will usually weaken to under one mG at 2'-3' out, though makes and models vary.

Net current

If a pair of conductors are carrying unequal currents there will be an imbalance, called a net current. In a typical two-conductor circuit, if the "hot" is carrying 10 amps and the "neutral" (GC) 4 amps there will be a 6 amp net current. This will act just like a 6 amp current in a single conductor. The magnetic field will be 12 mG at 1 meter (and 6 mG at 2 meters, 3mG at 4 meters, etc.)

Since GFIs (ground fault circuit interrupters) are designed to trip from a net current of five milliamps the net currents which cause detectable magnetic fields in buildings would have tripped any GFCIs instantly if they had been installed. This is another reason for extending the use of GFCIs and GFIs. (A GFCI, installed as a breaker, protects the entire circuit).

A magnetic field generated by a net current in a line source, whether it is from an unbalanced wiring circuit or from a water pipe weakens directly with the distance from the source. To detect a net current, take a measurement with a gaussmeter at some distance from the suspected source (say 2' from a wall). Then double that distance (to 4') and take another measurement. If the second measurement is about one half of the first, you are measuring a net current field from a line source.

If the second measurement is nearer to one fourth of the first measurement, you are measuring a balanced current field from a line source (typical of balanced power lines). The field weakens with the square of the distance, $1/r^2$.

If the second measurement is closer to one eighth of the first measurement, you are measuring a field from a coil-type source, which usually means a motor or transformer. The field weakens with the cube of the distance, $1/r^3$.

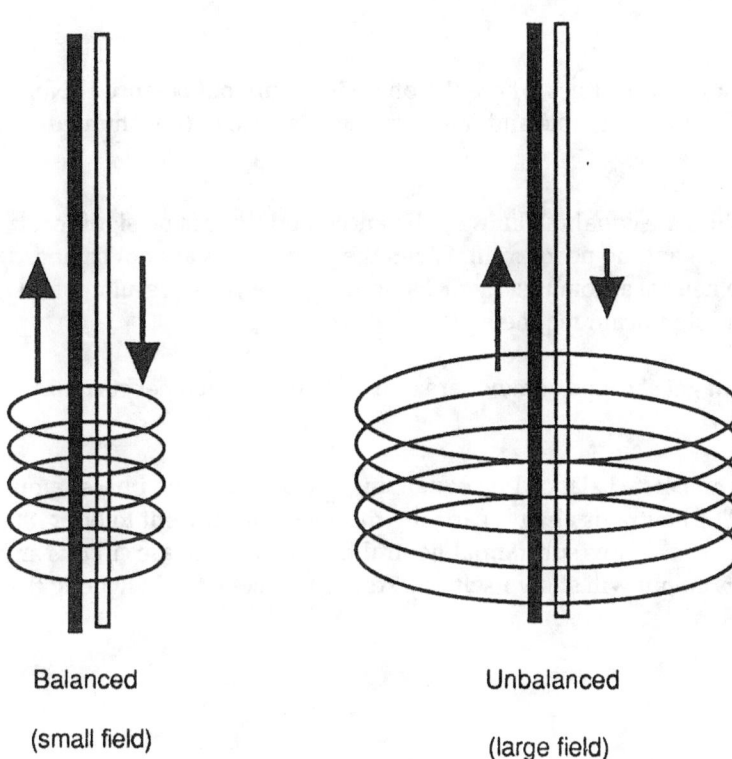

Balanced

(small field)

Unbalanced

(large field)

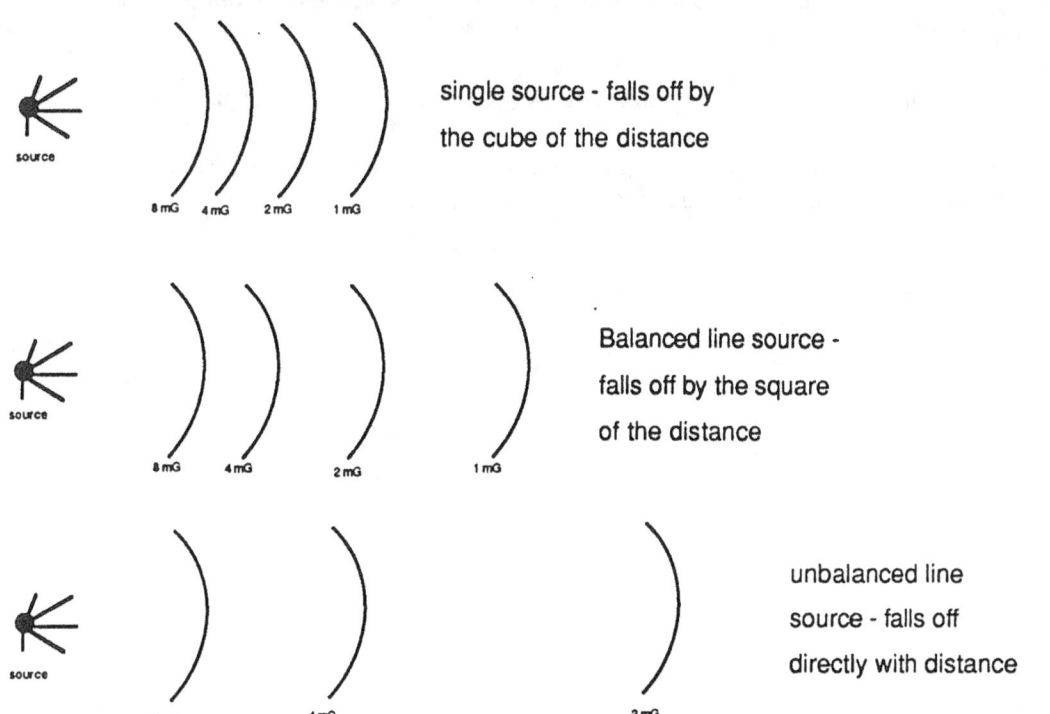

single source - falls off by the cube of the distance

8 mG 4 mG 2 mG 1 mG

Balanced line source - falls off by the square of the distance

8 mG 4 mG 2 mG 1 mG

unbalanced line source - falls off directly with distance

8 mG 4 mG 2 mG 1 mG

Three phase lines

Is there a net current on two or three phase lines where the phase loads are not balanced? No, as long as the associated neutral runs with the cable and it is carrying its full load (not shunted to a piping system, etc.).

When the loads on three phase lines are equal the lines are balanced and there is no significant magnetic field away from the lines - and no net current. When the phase loads are not balanced, as is the usual case in reality, the neutral automatically picks up the phase vector resultant and there is still no net current and no significant magnetic field.

But let some of the neutral current get shunted somewhere else and you've got a net current. More on this later.

A two phase service (120/208V) *cannot* be balanced by equalizing loads on the two phases alone since the phases are 120° apart. The neutral is always carrying the balancing current to bring the vector resultant to zero. Since there is always substantial neutral flowing on 2-phase circuits any shunting of neutral away from its circuit will show itself as a strong magnetic field all along the circuit path.

Single phase service

The usual residential service (120/240V) is "single phase", which means that one of the three phases from the primary distribution line goes to power the local transformer. The two hot legs coming out of the transformer are 180° out of phase with each other, so if their loads are equal there will be no net current. The neutral will carry the resultant if the loads are not equal, so there will still be no net current. Each 120V hot conductor is 180° out of phase with its accompanying GC; hence no net current. But put a hot conductor from one leg together with a GC from a circuit powered by the other hot leg (by mistake) and the fields of each will *add* together instead of canceling. This is because the GC from one circuit is *in* phase with the hot from the other circuit so the magnetic fields add.

As we will see later this is one more reason why GCs should stay with their paired hots.

Sometimes electricians call the two hot legs of the 120/240V service "phases". This is OK as long as we know what we mean. They are indeed 180° out of phase.

Chapter 4

NET CURRENT ON CIRCUITS: HOW IT GETS THERE

Since we know that magnetic field strength is directly proportional to net current all we have to do is find out how net current is created and eliminate it. Actually, that is how it works. Finding the exact location of the current shunt may take some detective work.

Remember that net current on a circuit means a violation of NEC (Exceptions comprise conductors designed to produce heat). Any net current over 5 milliamps would trip a GFCI.

Net current from neutral-to-ground bonding in subpanels

Why are neutral buses in subpanels insulated from the case?

Neutral bus bars in subpanels are insulated for a reason. Neutral current is not to be allowed access to equipment grounding conductors until it reaches the supply side of the main service disconnect, where neutral is bonded to ground.

> From **NEC Article 250.24 (A)(5) Load-Side Grounding Connections.** A grounding connection shall not be made to any grounded circuit conductor on the load side of the service disconnecting means except as otherwise permitted in this article.

For non-electricians this means no grounding of the neutral or GCs other than at the service entrance where the main cut-off switch is located for the building.

And yet it is surprisingly common to find neutral bonded to ground in subpanels by one or more of the following means:

• A service entrance panel is used for a subpanel and so there is no insulated neutral bus and the installer does not add one.

• The bonding screw on the insulated neutral bus is turned all the way in, grounding the bus to the case. (The bonding screw is to be used only when the panel is used as a service entrance panel).

• One or more equipment grounding conductors are connected to the neutral bus.

• The neutral bus is deliberately bonded to the can (box) with a strap or heavy wire.

• A neutral conductor or GC is connected to a grounding bus instead of to the neutral bus.

• A breaker panel is mistakenly assumed to be the main disconnect. Neutral current may get on grounding paths inside buildings when the electrician assumes that a breaker panel is the service entrance point when it is actually not. A building may have the main disconnect on the outside of the building and the breaker panel somewhere inside the building at some distance from the entrance point at a more convenient location. The grounding of the neutral should have been done on the supply side of the main disconnect outside the building. But the electrician, treating the breaker panel as the service entrance point, grounds the neutral bus in the panel. This may put neutral current on the pipes and conduits in the building where it can take convoluted paths before connecting back to the correct bonding point where it has access to the system grounded neutral leading back to the transformer to complete the circuit.

I am reminded of a breaker panel in the basement of a house in Berkeley, CA, where it was conveniently located at the bottom of the basement stairs but about 20 feet from the service entrance point. The neutral was bonded to a copper water pipe which carried the current along the basement ceiling where it touched a gas pipe which then carried current in a loop and dumped it back on the water pipe at another location. All this produced elevated magnetic fields in several of the first floor rooms. I clamped an ammeter around the water supply pipe outside the house and found no current to speak of. The circulation of neutral in the pipes was entirely internal to the house and simply provided a path for the neutral from the breaker panel back to the service entrance point where it was bonded to the service neutral. There was no net current on the service drop or the water supply pipe. The water pipe problem was solved when an electrician removed the prohibited neutral/ground bond at this panel. The neutral/ground bond already existed at the main disconnect outside the house.

See **Figure 3** for diagrams of correct and incorrectly wired subpanels.

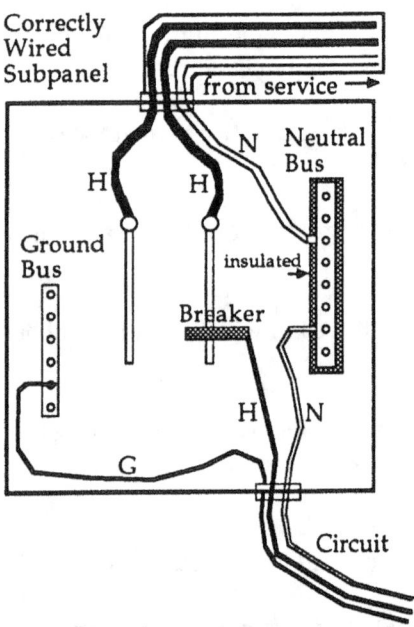 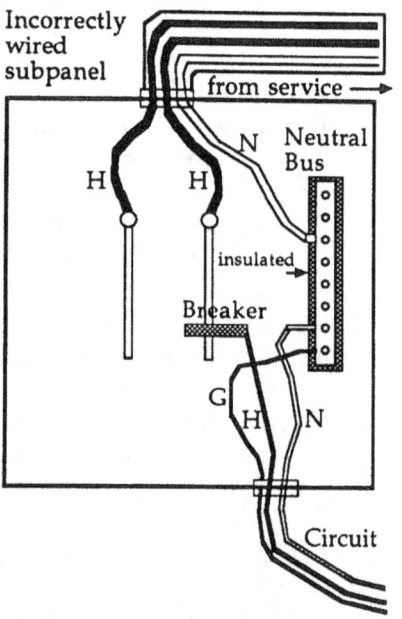

Figure 3. Schematics of correctly and incorrectly wired subpanels. Equipment grounding conductors must have their own bus. Neutral bus is insulated.

The consequences of incorrect neutral/ground bonding

When the neutral return current from the various circuits is given access to equipment grounding paths it splits and travels on both types of conductors according to the relative impedances of the conductors.

Impedance is composed of resistance and reactance. The cross section size, length and material of the conductor determine resistance, as well as the tightness of any connections. Reactance is increased by the separation of the conductors.

Remember that neutral current is the return flow which is headed back to the transformer supplying the site. It is attempting to complete its circuit, not "get to ground". Metallic pathways often present a low-impedance alternate path back towards the transformer for the neutral current. The ground itself presents a path but a poor path because of high contact resistance between the rod and the earth. (More on this later).

If neutral current has been shunted to equipment grounding paths and the wiring is in conduits the conduits with their large cross sectional areas present a lower impedance than the neutral wires. If the conduits or exposed equipment grounding conductors happen to touch a water or gas pipe with its even larger cross-sectional area, the impedance is even lower and more of the neutral is bled off to the pipes.

Figure 4 shows the effect on a high school classroom of bonding neutral to ground at a subpanel, and also the effect of correcting that error. See also **Figure 17** for a contour map of these fields.

I have found neutral current on every metallic pathway in buildings. It will spread through the metal framework of a hung ceiling, through heating ductwork, travel through cast-iron radiators in classrooms, circle aluminum window frames, spread through metal lathe, re-concentrate on gas or water pipes, etc.. Sprinkler pipes present an easy path for shunted neutral. Building steel may also become involved.

The 1993 NEC Handbook warns of the hazards created by grounding the neutral somewhere in the building (load side). On page 181 the authors describe the situation in which the neutral has become open on the line (supply) side of the grounding/bonding point. This would result in a condition where both the neutral and the metallic surfaces it has been connected to develop a voltage in relation to ground. This in turn could cause arcing in concealed spaces and could lead to "severe shock hazard" particularly for someone working in the building where they might disconnect a neutral or a pipe carrying neutral.

Figure 5 shows a typical sequence of net currents caused by an accidental (carpenter-caused) shorting of neutral to vent pipe in a ceiling. Neutral travels by contact from vent pipe to cold water pipe to hot water pipe to hot water heater to equipment grounding conductor back to the service entrance panel.

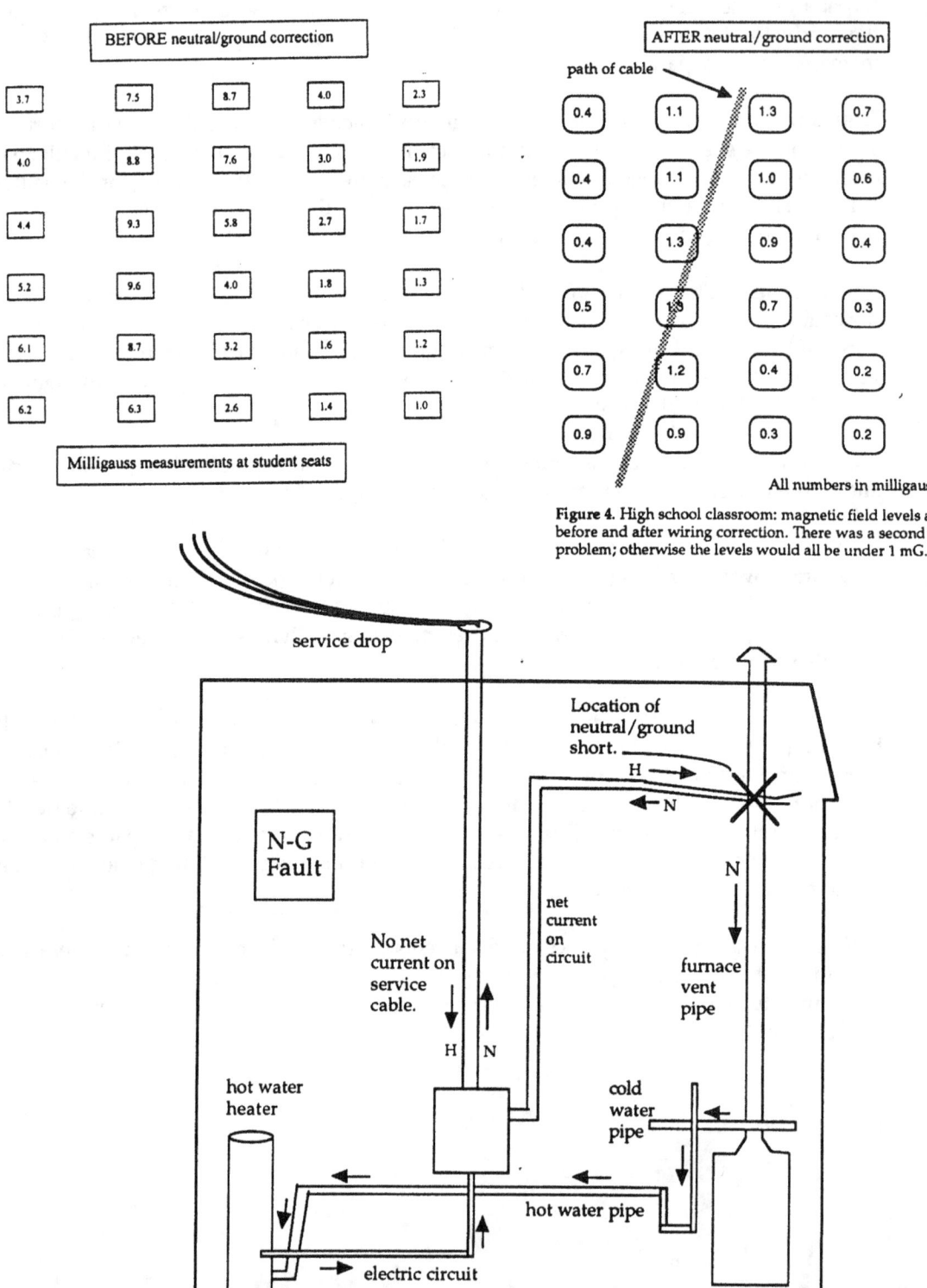

BEFORE neutral/ground correction

3.7	7.5	8.7	4.0	2.3
4.0	8.8	7.6	3.0	1.9
4.4	9.3	5.8	2.7	1.7
5.2	9.6	4.0	1.8	1.3
6.1	8.7	3.2	1.6	1.2
6.2	6.3	2.6	1.4	1.0

Milligauss measurements at student seats

AFTER neutral/ground correction

path of cable

0.4	1.1	1.3	0.7	0.2
0.4	1.1	1.0	0.6	0.2
0.4	1.3	0.9	0.4	0.3
0.5	1.3	0.7	0.3	0.5
0.7	1.2	0.4	0.2	0.5
0.9	0.9	0.3	0.2	0.5

All numbers in milligauss (mG)

Figure 4. High school classroom: magnetic field levels at seats before and after wiring correction. There was a second wiring problem; otherwise the levels would all be under 1 mG.

Figure 5. Actual path of neutral current inside a house in Elko, Nevada as a result of neutral shorted to a vent pipe which had been recently installed by cutting through the ceiling. Neutral was transfered through contact to a cold water pipe which touched a hot water pipe which transfered the neutral to the grounding conductor of the circuit and hence back to the entrance panel. This current produced elevated magnetic fields throughout the house. A temporary fix eliminated the current: a piece of plastic wedged between the vent pipe and the cold water pipe. N/G short to be located.

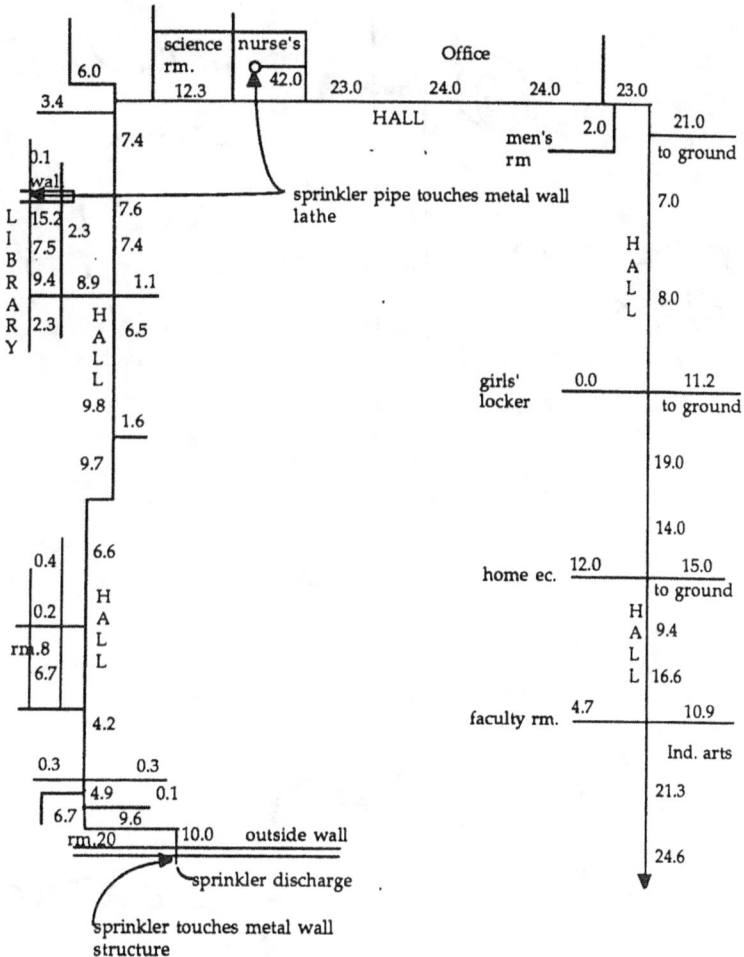

Figure 6. High School - diagram of sprinkler system with AC magnetic field mG levels, recorded before wiring corrections.

Figure 6 shows the magnetic fields coming from a sprinkler system in a high school as a result of multiple N/G shorts, including subpanel N/Gs. The currents traveling in the sprinkler pipes causing these fields have their counterparts as net currents in the circuits from which neutral current was shunted away.

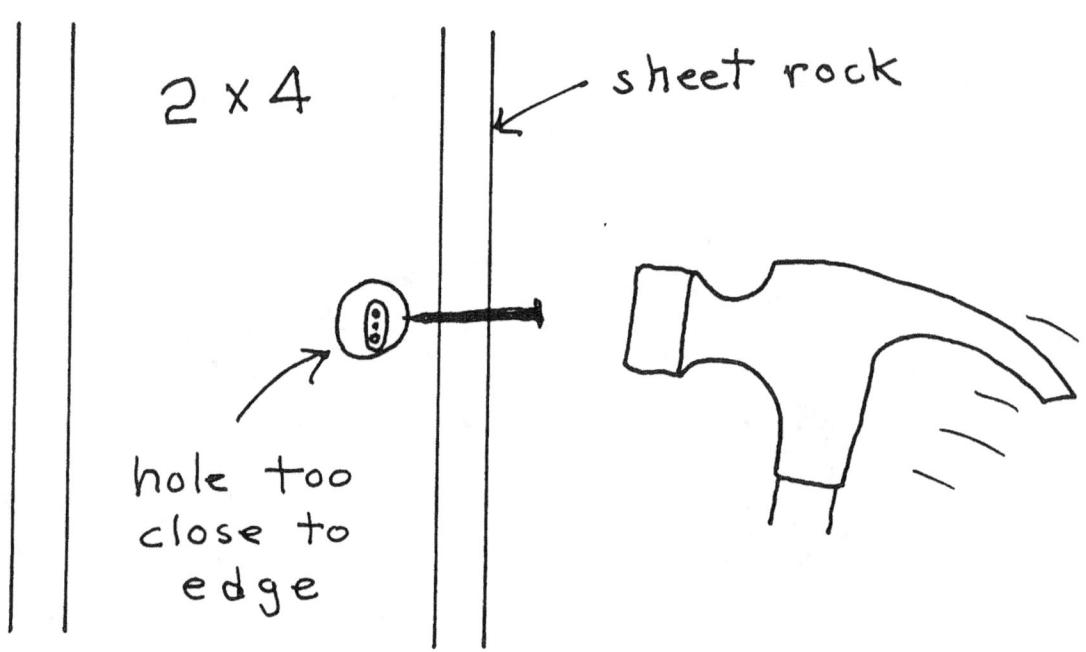

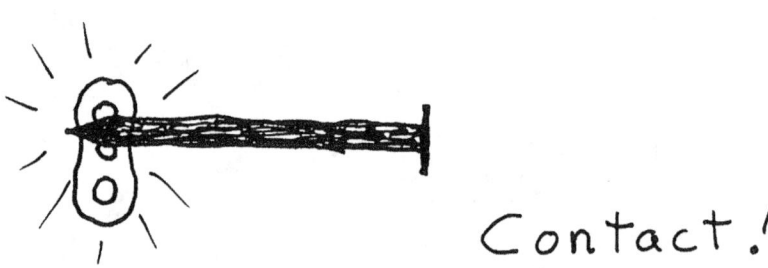

Not all neutral-to-ground connections are deliberate. Some are accidental or caused by a carpenter's nail. See Case Histories chapter.

Two or more fields where there should be none

When neutral return current is shunted to equipment grounding paths two or more net currents are created.

First is the net current on the feeder circuit going to the subpanel, which has been robbed of some of its balancing neutral. Second is the neutral current traveling on equipment grounding paths, paths which typically diverge from the circuit carrying the neutral conductor.

Fields from neutral-to-neutral connections

Now we get to perhaps the most common cause of elevated magnetic fields in larger buildings such as schools and office buildings.

It is a Code violation but is so commonly practiced by some electricians that they are not aware that there is anything wrong with doing it. I am speaking of the practice of wiring-nutting together all the neutrals in a junction box even though they may be from different *branch* circuits.

First let's see why this is a violation of Code, and after that how it creates high magnetic fields.

NEC **Article 300.3** [2002] reads:

(B) Conductors of the Same Circuit. All conductors of the same circuit and, where used, the grounded conductor [neutral] and all equipment grounding conductors and bonding conductors shall be contained within the same raceway, auxilliary gutter, cable tray, trench, cable or cord, unless otherwise permitted in accordance with 300.3(B)(1) through (4).

Looking at (1), we see that large conductors of size 1/0 AWG or larger can be paralleled according to Article 310.4, but this does not apply to ordinary circuits.

(2) applies to old wiring where there is no equipment grounding conductor. One can add one by following Article 250-130 closely.

(3) Now we get into the bowels of the 2002 AND 2005 NEC. This section says that conductors in non-metallic sheaths (NM cable) shall comply with 300.20(B). So we go there.

300.20(B): **Individual Conductors.** Where a single conductor carrying alternating current passes through metal with magnetic properties, the inductive effect shall be minimized by (1) cutting slots in the metal between the individual holes through which the individual conductors pass or (2) passing all the conductors in the circuit through an insulating wall sufficiently large for all the conductors of the circuit.

Confusing, no? It means that you are going to have to go around cutting slots in ferrous junction boxes or breaker cabinets and thus invalidating their UP approval if you start connecting your neutrals indiscriminately, or if you try to use 2-conductor travelers in 3-way wiring. I think it would be adviseable to ignore the exception for NM cable and keep all conductors and their currents together in the same cable. Remember that the NEC specifies minimum requirements for safety and there is nothing wrong with doing more than minimum.

The result of violating Article 300.3(B) is the creation of net current, and once again there will be two paths with equal net current.

How does wire-nutting together GCs (called neutrals) from different branch circuits violate this article? By connecting the neutral or GC from one branch circuit to another neutral or GC from another branch circuit, you make the neutral in the second circuit a part of the first circuit, and vice versa. You now have a conductor carrying current running separately from the other conductors of the circuit, in violation of 300-3(B). Likewise neutral current from the second circuit will split and some of it will now flow back through the first circuit. These neutrals or GCs constitute "parallel wiring of conductors" which is also prohibited by **Article 310-4**. There is no exception for NM cable ("romex") here.

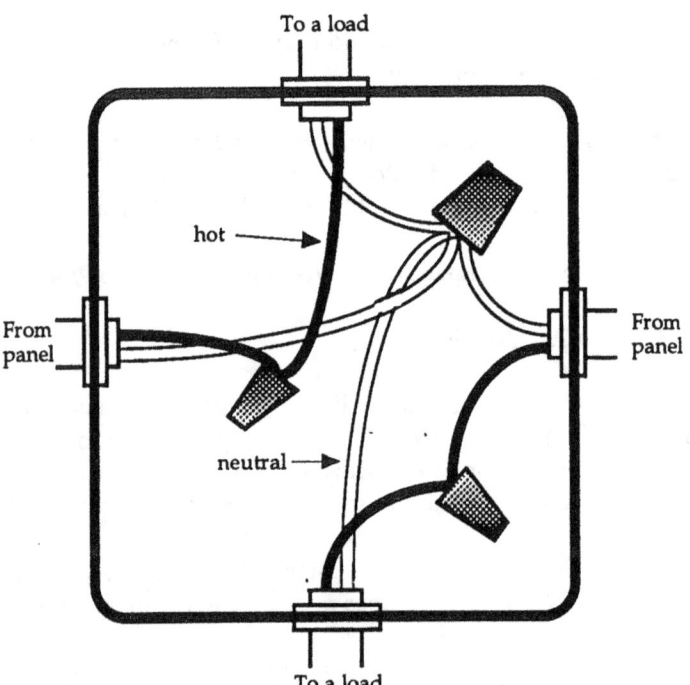

To a load

hot →

From panel

From panel

neutral →

To a load

Figure 7. Incorrectly wired junction box where neutrals from two branch circuits are joined, allowing neutral from either load to run on both circuits in parallel paths in different cables or conduits. Violation of NEC Article 300-3(b) and Article 310-4.

I ran my interpretation of Article 300.3(B), which to me was clear by that time, past the two Instructors at a ICBO-IAEI workshop on wiring and grounding multiple family dwellings, held for electrical inspectors. It took the two experienced instructors some time to consider the meaning of the article relative to paralleled GCs but they concluded that my interpretation was clearly correct. Local electrical inspectors may also need time to clarify the meaning of this article in their minds.

The second violation involved in the connecting of neutrals or GCs from different branch circuits is spelled out in **Article 310-4. Conductors in Parallel.** It permits only conductors of size No. 1/0 and larger to be run in parallel, and defines in parallel as "electrically joined at both ends to form a single conductor". This would apply to GCs joined at one end by a wire nut and the other end by a common neutral bus. The exceptions to this article relate to conductors carrying control power and to circuits carrying 360 Hz and higher, but even then they must run in the same raceway or cable.

What is the reasoning behind the NEC ruling that this is a violation? Separating conductors in a circuit causes increased impedance due to increased reactance in that circuit. Increased impedance results in heating of the conductors as well as metallic conduits they may be traveling in. Increased impedance means less current travels in the circuit. This in turn means a breaker is less likely to trip when a fault occurs, or will open too slowly.

There is also a shock hazard involved which is potentially lethal. Suppose an electrician is called in to add a circuit at a certain point in a building. He finds a convenient junction box where he can tap into a circuit with adequate capacity. He opens the breaker at the panel on the circuit that supplies that box so he can add the new circuit safely. He unscrews the wire nut on the hots and adds the new hot. No problem. He unscrews the wire nut on the GCs and in the process touches the return GC or the grounded box with one hand and the existing load GC with the other hand. He receives a 120V shock which could in some circumstances be lethal. Why? Because further out in the circuit someone has wire-nutted the GC from another circuit together with the GC from this circuit and it is carrying part of the return current from whatever load is on the other circuit. This hazard will be mentioned again in the trouble shooting sections.

Working with electricians I have more than once seen a look of surprise when they got a hefty spark from a GC they were disconnecting that was supposed to be dead.

There is always something wrong when large magnetic fields are detected in building areas. NEC recognizes one aspect of this: increased impedance leading to heating and tripping problems. Transformer and motor designers recognize this as a symptom of a poorly designed device which has failed to contain the magnetic field which is supposed to transfer energy within the device, not outside it. Those who are following the health concerns see a large magnetic field as intruding into biological space when it should be contained within the circuit or electrical device. Lawyers use the term "trespass" as well as other legal categories implying liability.

Why do many electricians bunch all the GCs together in any box they are working on? The reason goes something like this: first, joining GCs in a box is usually correct and necessary. The violation involves the less common case of a junction box being used by two different branch circuits. In this case the reasoning may go like this: "Neutrals are neutrals. They are all connected together at the neutral bus, so why not here?" The same reasoning could be given for connecting neutrals and grounding conductors, since all are bonded together at the service entrance. But the Code goes into great detail to specify where GCs may be joined and where GCs should be bonded to ground.

There are other ways the GCs from separate circuits can be incorrectly joined. One unusual way I have seen is this: a connecting tab on the side of a duplex receptacle is left intact. The receptacle is used for two circuits. The tab between the hots is correctly broken off to keep the hots separate, but the tab connecting the GCs is left intact resulting in a shunting of many amps of net current from one circuit to the other, causing major fields in the houses. (See **Case Histories** chapter).

Figure 8. Receptacle with break-off tabs when
separate circuits are feeding each end.

Incorrect 3-way switch wiring

I was called in by a California developer to measure the magnetic fields from a major transmission line which ran through his new development. In order to measure the field over a 24 hour period I set up a gaussmeter and data logger in the kitchen of a new house next to the lines.

When I turned on the instrument I was surprised by a high reading of about 12 mG whereas the power line outside the house was contributing only about 1 mG. When the kitchen lights were turned off, the field dropped to about 1 mG, which was from the power line.

I traced the strong field to a three-way circuit supplying the kitchen lights. The developer was not happy with this situation and set up a meeting between myself and the electrical contractor who had wired over 100 houses in the development. Over the phone he claimed all his wiring was to Code. I replied that if it was up to Code, then the strong magnetic field would not be there. He faxed me a diagram from an electrical book showing how he wired 3-way switches. I replied that if the circuit were wired the way the book showed there would be no strong field.

We met with the electricians at the house. Though they were reluctant to admit it, it turned out they had used a two-wire traveler where they should have used a three-wire. In this way they picked up the hot at one switch and used the GC at the other switch instead of having the GC return in the third wire to the box supplying the circuit.

The contractor asked for help in figuring out how he could balance out the magnetic field without cutting into the plaster to pull out his romex, but unfortunately there was no way. Perhaps he was also thinking of the other 100 or so houses that his men had wired in the development.

A contractor does not necessarily know what his men are doing in the field. A contractor with a gaussmeter will be able to tell instantly if certain errors have been made as will an electrical inspector with a gaussmeter, once the power is connected.

Correctly wired 3-way, 4-way, etc. switches done according to diagrams in wiring books will not produce unusual magnetic fields, as currents are always paired. But using 2-wire where 3-wire is called for violates 300.3(B). (Unless you are using NM cable and you want to use the exception and cut slots in any metal the separated conductors pass through, including fluorescent fixtures and junction boxes etc.) Your GC is not returning with your hot. Just remember, use 3-wire between switches.

Chapter 5

MAGNETIC FIELDS DUE TO SEPARATION OF CONDUCTORS

Power lines

The reason power line magnetic fields affect such large areas is only partly due to the amount of amps they are carrying. It is also a function of separation of conductors. The old way of spreading out secondary distribution lines horizontally and using widely spaced conductors for service drops caused relatively high magnetic fields. The present method of using spun secondaries and spun service drops reduces the magnetic field from balanced lines to negligible values.

For the same reason burial of transmission or distribution lines reduces magnetic fields not because the earth shields the fields (it does not) but because the conductors in underground cables are close together and self-canceling to a great degree, particularly when run together as part of a single cable.

There is a pole-mounted system which is much cheaper than direct burial but which brings the conductors within 8" of each other, causing much better cancellation than spread-out conductors. It is called the Hendrix Aerial Cable (trade name) and can be identified by the diamond shaped spacers on the lines which keep them spaced properly. These lines are insulated since they can carry up to 37 kV. There would be corona problems at this spacing without insulation. To appreciate the canceling effect of an 8 inch spacing compared to a 10 foot spacing it is enough to know that magnetic field strength is directly proportional to conductor spacing.

The **figure below** shows the canceling effect of 8" spacing compared with the conventional horizontal spacing as measured on a line which changes from one system to the other at a point where a State forest begins (the tight spacing reduces tree trimming to a minimum).

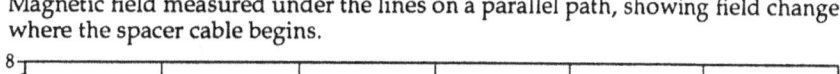

Magnetic field measured under the lines on a parallel path, showing field change where the spacer cable begins.

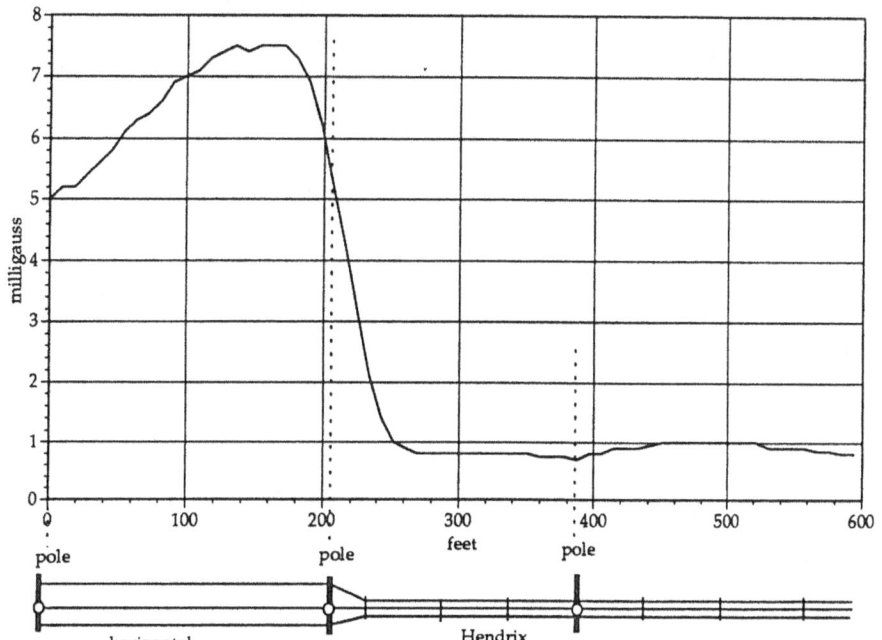

Knob-and-tube wiring

Inside older houses you may find knob-and-tube wiring, where the conductors are separated by a foot or more. The knob refers to the porcelain knob to which the wire is attached along a rafter, joist, etc. The tube is also porcelain where it goes through a stud, rafter, joist, etc. This wiring was generally discontinued in the early forties, but some buildings used it as late as the sixties. Knob-and-tube is no longer allowed by NEC but is grandfathered and can be left in place or added onto with contemporary wiring.

However if your client is concerned about magnetic fields there is no solution for knob-and-tube other that to replace it with contemporary cable. There may be seldom-inhabited areas of the house where it can be left in place but in living spaces it will have to be replaced if the customer wants to be free of the relatively high magnetic fields that knob-and-tube always produces.

The former use of knob-and-tube may help explain why older homes show a higher average magnetic field level than newer homes (see **Figure 21** in Appendix A). Our parents or grandparents may have grown up in higher residential magnetic fields than their contemporary children, despite our expanded use of electricity. There is no necessary correlation between a nation's electrical usage and individual magnetic field exposure, simply because magnetic field levels are determined more by conductor spacing than by load.

When source conditions remain more or less uniform, however, one may see a correlation. A study covering 8 Canadian provinces showed a small but statistically significant relationship between increased electric power consumption and risk of all childhood cancers combined and brain tumors in Canada as a whole.[1]

---

[1] EMF Health Report, Nov/Dec '94, p. 5f.

Chapter 6

MAGNETIC FIELDS DUE TO GROUNDING TO WATER PIPES

NEC requires that the metallic water pipes as well as other metallic systems that may be contacted by building inhabitants be bonded to the common neutral/ground point at the main service disconnect. This prevents a voltage developing on the metal surfaces that could be hazardous to any grounded person or beast. By tying in the pipes to the system grounded neutral any fault current which gets on the pipes will be pulled through the breaker with enough amperage (hopefully) to trip the breaker and open the circuit.

Internal gas pipes must be bonded in this way but since 1990 may not be used as grounding electrodes. Gas companies have taken care of this by inserting a dielectric coupler (insulator) at the meter. If not, the homeowner's gas service should be contacted.

Metallic water pipes however are considered a qualifying grounding electrode when in contact with the soil for 10 feet more. In practice they usually join the water main in the street and interconnect with all the other buildings in the area. Metallic water systems have become an uncontrolled part of the electrical distribution system, serving as conductors of neutral return current and providing a parallel competing path with the local service drops to the transformers which supply them.

They may be carrying a substantial percentage of the neutral return current between houses in the neighborhood. In a case where the utility's neutral connections at the pole or at the service drop are corroded and have increased impedance, most of the neutral return current may flow out the water pipe to the water main and then flow through the pipe into a neighboring house that may have newer service drops with good neutral connections. The result will be that the neighboring house's neutral service conductor will be carrying not only its own neutral current but some of the first house's also.

What does this mean for the magnetic field situation in the two houses? It all depends on where the water pipe comes into the house and where it is bonded to the electrical service. It also involves the location of the service drop relative to living areas in the houses. The figure below shows a situation where the pipe passes through the house before being bonded at the service entrance. Note that there is a net current field spreading out from the service drop which may run adjacent to a bedroom. An identical current on the water pipe will be affecting both the basement and the first floor areas.

Since net current is not affected by the spacing of conductors the fact that the service drops are spun has no effect on the magnetic field of the net current.

As an electrician you may ask what business is this of yours, since service connections are the utility's responsibility and you are required by Code to bond to the pipes. It is not your direct responsibility to eliminate this magnetic field; however you can inform the property owners that Code does allow the reduction of objectionable current by making certain changes. (**NEC Article 250.6 Objectionable Current over Grounding Conductors.**).

The most complete study of recommended ways to accomplish this was written be Fred Hartwell, senior Editor of **EC&M,** The Magazine Of Electrical Design, Construction, and Maintenance, March, 1993. He outlines several ways to eliminate the neutral current on water pipes which are in accord with Code.

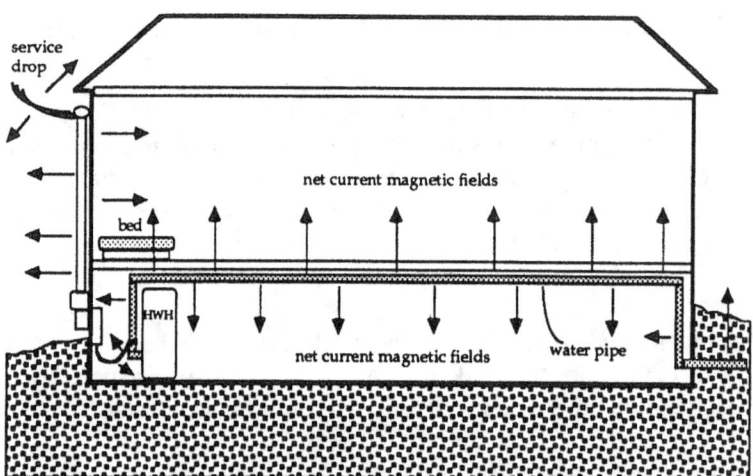

Figure 10. Path of net current magnetic fields when neutral return current splits at the service point due to required grounding to a metallic water system which enters at the other end of the house. The water pipe has a current and the service drop has the same net current due to a deficit of neutral.

I usually recommend the method which retains the water pipe as a grounding electrode by having a plumber insert a dielectric coupling in the water supply pipe at least 10' out from the building at some convenient place. If it is a suburban house, out near the curb is a good accessible spot. The 10' distance is specified in the NEC as representing the minimum amount of soil contact which qualifies the pipe as a grounding electrode. This 10' or more gives a good ground contact for the pipe in case of a lightning strike or power line accident but presents too much impedance to allow any significant low voltage neutral current to flow. If the ground is continually wet (sometimes from a sprinkler system) it would be better to insert 1' of plastic pipe as a dielectric to ensure the neutral current will be blocked. This may be necessary in areas where water pipes are buried 4' deep for frost protection. The water inside the pipe is a poor conductor at that voltage but ground water in the soil may contain conductive salts.

This leads us to discuss some points about what the local grounding electrode does and does not do.

Chapter 7

SOME MISCONCEPTIONS ABOUT GROUNDING

Soon after I began conducting magnetic field surveys on the island of Marthas Vineyard I was made aware that the subject of grounding can be an emotional issue with some electricians. I traced the emotional charge to misconceptions about the functions of the local grounding electrode and a misunderstanding of its role in electrical safety.

A simple question from an electrician whom I was working with put to a State official from the electrical licensing bureau about whether inserting a dielectric coupling in a water pipe 10' out from the building met with Code brought forth a tirade from the official which was so emotional that it cut off the possibility of discussion.

Some weeks later I talked to that official on the phone about grounding. I was trying to discover the reason for his agitation. I asked him what was the function of the local grounding electrode. He snapped back an answer: "The function of the local grounding electrode is to clear a fault current, pure and simple". Since I knew from reading Soares Book on Grounding (adopted as a text on grounding and bonding by the International Association of Electrical Inspectors) that this answer was exactly the opposite of the main point Soares was making in his book, I mentioned this to the official and stated that from my understanding the function of the local grounding electrode was not to clear a fault but to help keep the distribution system neutral at local ground potential and also help disperse a high voltage strike such as lightning or a high voltage line falling on a distribution line. He gradually backed down and admitted this was right, but the very fact that he gave the wrong interpretation initially explained why he had been overly concerned with any "tampering" with a water pipe ground. He was carrying around a false picture in his head of the safety situation.

The fact that two of Marthas Vineyard's towns had gone to plastic water systems without vociferous objections from the local electricians did not seem to lessen this official's sense of peril about inserting a dielectric spacer in a residential water pipe. Before I had the above phone conversation with the official he had contacted the newsletter, Codewatch to complain about this new "threat". The editors contacted me by phone and got my point of view. They printed a short account of the dispute in the next issue, but then they researched the issue and in a subsequent issue of Codewatch printed an article in which they recommended use of a dielectric coupling in the case of objectionable neutral current running on water service pipes.

But what does insure that a breaker will trip when a ground fault occurs? The 1993 NEC added a fine print notation (FP) at the beginning of **Article 250 - Grounding** designed to make this clear:

(FPN No. 1): Systems and circuit conductors are grounded to limit voltages due to lightning, line surges, or unintentional contact with higher voltage lines, and to stabilize the voltage to ground during normal operation. Equipment grounding conductors are bonded to the system grounded conductor to provide a low impedance path for fault current that will facilitate the operation of overcurrent devices under ground-fault conditions.

The last sentence was new to the 1993 edition and was added to try to clear up the common misconceptions about the function of local grounding electrodes, whether they are ground rods, building steel, rebars, Uffers, water pipes or whatever. Since the wording in the Code is very tight, let's decipher what it is saying.

It is making a distinction between the functions of grounding to the local grounding electrodes and bonding to the "system grounded conductor", which we have been calling the utility's neutral conductor.

Grounding locally is only for the purpose of:
1. Helping to disburse rare high voltage charges on the distribution system such as lightning as well as high voltage that might happen during a storm in which a high voltage line is blown over onto a low voltage distribution line and in that way high voltage makes its way into the building.
2. Helping to maintain the whole distribution system in a fixed relation to ground potential, so that the 120V-to-ground potential does not drift up to higher potential in relation to ground as the line gets farther from the substation.

The function of the local grounding equipment in relation to a lightning strike was brought home by the ICBO-IAEI workshop instructor who said that lightning goes the same place as an 800 LB gorilla: wherever it wants. All the local grounding system does is politely invite the lightning to come this way to earth. But the immense voltage and power involved is such that the actual path the lightning takes is not completely predictable and we can only follow some basic principles of trying to channel it and say we did our best.

The second function of the local grounding electrode of helping to keep the system at constant voltage in relation to ground can be understood in terms of the build up of voltage in an ungrounded line as it gets farther from the substation, but since this is not a book for utility engineers please refer to the bibliography for two good books on power line problems by O. C. Seevers. Each local electrode plays a part in maintaining a constant voltage potential in the system.

Does the earth connection clear a ground fault?

Now we can ask the question, why has not the clearing of a ground fault been included in the list of functions of the local grounding electrode? For non-electricians, "clearing a fault" means allowing sufficient amps from an accidental contact between a hot conductor and grounded metal to travel through the circuit breaker and trip the breaker, thus killing (opening) the circuit.

The reason is that the earth/electrode boundary has too much resistance to be relied on to allow enough current to flow to trip even a simple 15 Amp breaker. The impedance is too high. Let's do some simple math. First, how much impedance does a typical ground rod have to earth? It depends on the type of soil and its condition, but a university study of the ground rod resistance of 1,172 utility ground rods in Montana, Iowa and Wisconsin yielded an average resistance of 119 ohms! A 120V ground fault could push only 1 amp into the earth. Thus the ground rod would remain at 120V to earth, and anyone touching it would get the full voltage, depending on where they stood.

One utility reported the resistance of their ground rods varied from 40 ohms to 1150 ohms. 40 ohms would allow 3 amps at 120V. No breakers would trip, and touching the ground rod itself could give you a 120V shock depending on how close you stand to the rod.

Now the NEC states that if a single ground rod does not have a resistance of 25 ohms or less, it must be supplemented with another rod. The combination is not required to meet the 25 ohm limit or any other limit. Why? Because the system does not use the connection to earth to provide safety from ground faults in the building's wiring. Where, then, did the 25 ohm requirement come from if it is basically meaningless? One inquiry into the history of this requirement traces it back to a primary distribution power line spec for grounding. If this is the case it makes sense, as we are talking about 69kV or more instead of 120V. 69kV can push 2,760 amps through a 25 ohm rod, whereas 120V can only push 4.8 amps through,again not enough to trip a breaker.

The real safety connection

Safety, the clearing of faults, is accomplished by the solid connection of the equipment grounding conductors to the neutral bus at the service box, and the bonding of this bus to the service neutral, which takes it back to the transformer to complete the circuit.

This solid connection allows a surge of current to pass through the breaker, which trips the breaker and provides the safety. GFIs trip even faster and provide safety from shock, whereas the breakers trip in time to prevent fire but not necessarily fast enough to prevent heart fibrillation. To get a reasonably fast breaker response you need at least 5-6 times the breaker rating to trip with speed. In other words, for a 15 amp breaker you would want the system to pass at least 75 amps, which means a resistance of 1.6 ohms.

If you were relying on a ground rod alone, you would fry all day, since no breaker would trip.

So it is a solid connection to the electrical distribution system's grounded neutral conductor that has low enough impedance to allow sufficient fault current to flow to trip a breaker. "Pure and simple!"

Origins of the Grounding Safety Myth

So the idea that a connection to earth provides safety in a 120V system is pure myth. So why do so many people, including many electricians, electrical engineers, and even some physicists believe that connection to earth provides safety from faults in 120V systems? This myth cannot be found in the National Electrical Code nor in the authoritative books on grounding. I have found some of the sources of this pervasive misunderstanding.

- **Popular home-electrician books**

I checked at a local public library. There was one book on electrical wiring, very well illustrated and published by a well known company in collaboration with a tool manufacturer. I looked in the glossary and found:

"Grounding wire: A wire used in an electrical circuit to conduct current to the earth in the event of a short circuit."

False! The circuit grounding wire (equipment grounding conductor, or EGC) is there to conduct current from a ground fault (not a short circuit, which does not involve the grounding wire) to the neutral/ground bond at the service point panel, from where it travels back to the transformer in the service neutral conductor to complete the circuit. This low impedance path allows enough current to flow through the circuit breaker to trip it and kill the circuit. The connection to earth doesn't help at all. It is only there to help direct huge currents from lightning strikes, high voltage line surges, etc.

So why this myth about the earth? Another quote from that book may provide a clue:

"The earth has a unique ability to absorb the electrons of electrical current. In the event of a short circuit or overload any excess electricity will find its way along the grounding wire to the earth, where it becomes harmless."

This is a curiously child-like view of Mother Earth that has nothing to do with how electrical circuits behave. The book has evidently been written by a professional writer, not by an electrical expert. And it is being read by homeowners who have no reason to disbelieve these statements.

Let's look at the statement. First there is a picture of "excess" electrons needing somewhere to go. "Overload" is also used inappropriately, implying that too many electrons are in the circuit and need to be bled off. Then there is the picture of the earth <u>absorbing</u> electrons, making them "harmless". This is the "earth as a sponge" image which we will look at next.

The fact is that electricity travels in circuits, and only flows when the circuit is complete. The circuits in a building begin and end at the transformer. Electrons move out and return in the conductors. Where building wiring is concerned, the earth is not involved in the circuit except as an inconsequential parallel path which provides no safety. (There can be dangerous situations where it might be, as we will see later).

• Earth as a sponge

Electrical professionals as well as lay persons carry an internal image as to how electricity works. Since it is invisible we have to picture it somehow. Some of us have a feely image more than a visual one, but we have to have some simplified way of understanding and predicting. Scientists call this a "model".

I believe that many persons cary an image of the earth as an electrical sponge. As long as a structure is "grounded" to earth, all is well, since Mother Sponge will absorb the electrons and render them harmless. I guess harmless means they stop moving. Current over!

But is this realistic? Since electricity travels in complete circuits, the only part the earth can play is as a conductor in the circuit. Since buildings are all grounded to earth it is always an available parallel path (in the US) between the service entrance point and the transformer, which is also connected to earth. But if one measures how much current is flowing back by way of the grounding conductor through the earth, most ammeters will show zero, depending on how sensitive they are. This is with typical earth conditions. There are exceptions.

Since earth can only play the role of a conductor, why is it thought of as a sponge? One explanation has occurred to me. When lightning strikes the earth the current spreads out and the path can be followed for a ways. So one might say the earth absorbed the lightning current, like a sponge. But we are only seeing a portion of the gross circuits of charges that build up and equalize between the atmosphere and earth. The electrons don't just stop; they move along in response to the forces building up charges between earth and cloud. These are circuits being completed, albeit suddenly and dramatically. And of course some of the strikes are actually upward, from earth to cloud, so how spongey is that?

• Earth as a good conductor

The idea (model) of earth as a sponge makes no sense. The idea of earth as a conductor is valid, however, and must be addressed. It is a point which has caused confusion. The reason grounding electrode systems have such high impedance is due to resistance at the interface between the electrode and earth, both at the building and again at the transformer grounding rod.

The earth itself has been shown to have negligible resistance. This was a surprise to me when I learned that there were studies which showed this to be the case. The explanation seems to be that even though any constricted earth path has high resistance, the paths available are almost infinite, so you have an almost infinite number of parallel paths, electrically speaking.

I was also surprised to learn that in some parts of Alaska, Utah, Australia and several other countries the earth is still being used as the sole neutral current return path of the primary distribution circuit. The system is called SWER (single wire earth return). It is used in remote regions to save money on the cost of installation by running only a hot conductor to the remote transformer. How can this work?

If earth resistance itself is negligible then you just have to overcome the resistance between electrode and earth. This can be done in two ways: construct a system of electrodes in parallel to lower total impedance, and keep the amperage low enough to avoid dangerous voltage buildup at the remote ground. This is typically achieved by limiting total primary neutral return to 8 amps and using enough ground rods or other electrodes at the base transformer to achieve 2 ohms impedance. This is for the primary distribution line, typically 12.7kV. The secon dary lines have their neutral and are conventionally connected. (See www.ruralpower.org/swer)

So to summarize the function of earth as a conductor, the fact that it carries almost no current between a building and its transformer is due to the resistance between electrode and earth. The earth itself has negligible resistance because of the almost infinite paths the current can flow in. One can only use the earth in this way by constructing multiple or specially engineered electrodes.

Multiple Grounding Electrodes

The use of multiple grounding electrodes is the very situation we have in any standard primary distribution line (wye). It is referred to as a "multi-grounded system". The transformers and every few power poles have ground rods connected to the neutral conductor. All are in parallel with the primary neutral conductor on the poles, which return current to the substation. (We use DC terms when speaking of an AC system, but it works in general).

Because each rod contributes a small percentage of the returning neutral current, together they allow a typical 30% or more of the current to flow back to the substation in the earth. Studies show that the path through the earth is not infinite but in a virtual "tube" underneath the power poles.

Stray Current

The possibility that some of this neutral current might find an easy path through a dairy farm is the basis for some of the concern about "stray voltage" or "stray current" effects on milk production. For those who are new tho this, it is well to know that hoofed animals such as cows and horses are a lot more sensitive to small voltages than humans. The use of the word "stray" also reveals a cloudy mental model of electrical flow. Neutral current does not "stray", but follows all available paths to complete the circuit back to the transformer or substation.

Stuart Maurer's Study

A professor of electrical engineering got a grant to study the currents flowing on power line grounds along a primary distribution line. He found that the highest neutral currents flowed to earth at the end of the line, lessening as one measures nearer the midpoint of the line, after which neutral current starts returning from earth through the rods to the neutral conductor. The maximum return is through the substation grounding grid.

How much current travels through the earth is purely a matter of relative impedances. A ballpark impedance of 0.55 ohms per thousand feet of primary neutral conductor is competing with the paralleled ground rods.

Though this return current in the earth usually flows under the power lines, any metallic piping or rail systems which cross under or parallel the lines will cause some of this current to travel in the metal as another parallel path back to the substation. A line which loops around an area will allow possible short cuts for the current away from the power line path. The possibilities for voltage in facilities which border the lines or are in the middle of loops is there and should be kept in mind when investigating "stray voltage" occurrences.

Summary

Earth itself is a good conductor of electricity because it represents almost infinite parallel paths. But to get to earth the current has to cross the impedance barrier between the grounding electrode and the soil. It has a second barrier when crossing back to a ground rod at the transformer.

A single electrical service at a building typically has a minimal connection to earth, so in the case of a ground fault not enough current will flow to earth to allow a breaker to trip. Hence the hazard remains.

A primary power distribution line is in a different situation. It has multiple grounding points throughout the system all working in parallel. Hence the impedance is lowered and the earth becomes a second path for neutral current returning to the substation transformer.

A local grounding electrode can carry high voltage currents such as from lightning or surges from crossed lines during a storm.

Use of "ground" when the real meaning is "bond"

Perhaps the main cause of misunderstandings about the limited functions of connecting to earth is the misuse of the word "ground" in electrical publications. The connection which allows a breaker to trip in ground fault situations is the <u>bond</u> between the grounding bus and the service neutral at the building service entrance. This allows fault current to return easily to the transformer so that sufficient current will flow through the circuit breaker to trip it open. The earth is not involved. Hence to speak of "safety grounding" to earth is a false conception.

Unfortunately, all through the official electrical books the word "ground" is used when the meaning is to bond or to lead to this bond. Most lay persons, physicists and apprentice electricians are going to assume that ground means earth.

To remedy this situation electrical experts such as Mike Holt have been trying to get the National Electrical Code panels to change the word "ground" to "bond" when the meaning intended is indeed to bond. This effort is finally producing some results. The 2005 NEC has changed much of the wording, so that Article 250 has been changed from "Grounding" to "Grounding and Bonding".

Vested interests

I have been surprised to read occasional articles in reputable electrical journals about achieving "electrical safety" through lower-impedance grounding electrode systems. These articles are written by engineers employed by companies which install these systems. Though they may be useful or even essential for supplying a good common ground voltage reference for telecommunications equipment and other electronics, the vicarious inclusion of "safety" as a reason for installing the system simply helps to perpetuate the myth that connection to earth provides safety in relation to building voltages. And of course it helps to sell their product.

Using water service pipes as grounding electrodes

Formerly water pipes were all metal. Since they were in good contact with the earth it made sense for the NEC to require that they be used along with other grounding electrodes for the earth connection to disperse lightning strikes and other high voltage events.

This makes sense, but one consequence was evidently not anticipated. If the water system is shared with other buildings in the neighborhood, a parallel path for neutral return current is set up that involves neighboring buildings. Why? Because the other buildings also ground to the water system and may be supplied by the same transformer. This allows neutral current to travel from the neutral bus in one building down the grounding electrode conductor to the water pipe, through the water pipe to the neighbor's house, up his grounding electrode conductor to his neutral bus and on back to the transformer through the service neutral conductor.

So now the neutral current produced in one building has two paths to return to the transformer. It moves through the service neutral conductor and also through the water pipes to the neighbor's service neutral and back to the same transformer. This results in a completely unplanned and unregulated parallel current situation. The NEC goes to great pains to prohibit parallel current paths. Where the large size of conductors makes it desirable to use two per phase to carry high currents, they must be clustered together with the other phases and sized exactly the same (NEC 310-4). And in the case of conductors smaller than 1/0 AWG, paralleling is prohibited.

The water pipe paralleling violates these provisions but since it is outside the service point the NEC does not apply.

Why is paralleling prohibited or carefully regulated? Because otherwise it sets up net currents due to unequal currents traveling in the conductors. And what do net currents do? They induce currents in any condcutive material neaby. This results in a heat rise, and that is what the NEC is concerned with, since all their regulations are designed to keep the heat range within the limits of safety for fire concerns as well as for the premature degrading of insulation and eventual failure.

This book is focused on EMF effects, or the production of elevated magnetic fields. So we are concerned not only with the heating effects of currents induced by the net current magnetic fields, but other effects such as interference with electronic instruments and the health effects now verified and continually investigated by the health research community.

The water supply system as an "excellent" ground

There may still be a lingering safety concern about inserting an insulating fitting into the water supply pipe. The reasoning goes like this: "The metallic neighborhood water system is an excellent ground, so why loose this additional safety factor just because you are worried about an unproved cancer effect?"

But is it indeed an excellent ground? The definition of even an "adequate" ground would be that it enables a breaker to open during a ground fault condition. Here is what Soares Book on Grounding has to say about that:

> If we were to assume the unlikely but nevertheless possible condition that the neutral at the transformer was grounded to the same water pipe system then the resistance of the fault path would be appreciably decreased. [compared to a simple earth connection] Because of the wide separation between the service conductor and the water pipe, the reactance and thus the impedance of the fault circuit would remain high. The probabilities are that the fault-current would not reach a high enough value to operate the overcurrent device. Again fire and damage to equipment would continue until the circuit was manually opened. To improve the safety of such a system, a low-impedance path must be provided to pass enough current to clear the circuit through the overcurrent devices. [This refers to the neutral conductor in the service drop - Author]. P. 73, 4th Edition.

Since there is so much misunderstanding about local grounding, at the risk of belaboring a point I quote again from Soares Grounding Electrical Systems for Safety, 3rd Edition, IAEI.

> ...no grounding electrode no matter how low its resistance, can ever be depended upon to clear a ground fault on any distribution system of less than 1000 volts. p.174.

So the only effective "ground" when we are talking about clearing a fault is the **system ground,** which means a solid connection to the incoming service grounded neutral. This gives access to multiple grounding electrodes as well as the crucial direct connection back to the transformer to complete the circuit.

I realize that there are some well informed engineers who will argue that the water system is a better neutral ground path than is described by the experts cited above. There are some who say that if a building looses its neutral in an unusual accident in which the phases escape untouched, they would rather have the power stay on, using the water pipes as the return neutral. Otherwise there will be voltage differences on the two hot legs which may damage appliances in the building and cause some lights to blow out. This is a judgment call. Personally, given the undependable and completely accidental degree of impedance in the metallic water system in any particular instance, which may lead to prolonged ground fault conditions which result in fire hazard, I would rather see the clear and immediate effects of loss of the service neutral so that power can be shut down and the problem corrected. Imagine relying for any length of time on a circuitous path of neighborhood water pipes to complete the circuit to the transformer. And imagine the electrocution hazard to any water works employee who had to disconnect or cut one of those pipes.

Since one hears so many opinions about the relative impedance of water systems, it might be a good idea to switch from opinion to reality. The next time you see a current on a grounding electrode conductor or the pipe it is connected to you can take two simple measurements to determine the relative impedance of the pipe system VS the service neutral. First determine whether you are measuring neutral which is being produced from loads in the building (by opening the main switch if feasible and verifying no current on the neutral). Once you verify that no external neutral is coming through the service line, place a constant load on one 120V circuit in the building and clamp your ammeter around the grounding electrode conductor or pipe. Note the amperage. Then clamp around the service neutral conductor.

The ratio of the two currents will be the ratio of the impedances of the two paths. The amperage on the water system would have to equal the amperage on the neutral for the water system to compete with the neutral. A higher impedance on the neutral could also indicate connector corrosion which should be corrected by the utility, after which you can re measure.

If the building is in use and you cannot control the loads, these two measurements can still be made with two clamp-on ammeters simultaneously, since the neutral will fluctuate. If the fluctuations are more or less stable you can also use a single averaging ammeter like the Fluke 33 or else a clamp-on going to an averaging multimeter like the Fluke DMM 83.

A second reason given by some utility engineers for retaining the water system as part of their electrical distribution system is that it helps the neutral return current on their *primary* lines to return to the substation. When primary and secondary lines share a common neutral some of the neutral due to imbalances on the primary may flow through the secondary neutral system, flow into buildings on the service neutral and out through the water pipes to the water mains which may be bonded to the substation grounding grid.

If metallic water systems have been relied on for effective return of primary neutral in the past, the very fact that towns and cities as well as new housing developments are increasingly using non-conductive water mains should lead these engineers to get to work on revising their way of producing, regulating and returning neutral current on their systems. I understand that there is a lot of work being done on this problem by EPRI engineers, which includes looking at some European systems which are more electrically advanced than our own.

One promising solution to the water pipe problem which could be applied system-wide as well as individually was described in a poster at the November, 1994 DOE-EPRI EMF conference in Albuquerque. The system is called Net Current Control, or NCC as described by D.W. Fugate, ER&M; J.H. Cooper, PDC; and R.J. Lordan, EPRI. To quote: "The objective of this project was to investigate a specific means for reducing or eliminating these net currents without compromising the efficacy of the ground system in any way". Without changing the impedance of the grounding system, the device reduces net currents from amperes to milliamps.

How does it work? The authors would not reveal details pending their patent application; however, Spark Burmaster had a 12V model of his version of the system and it is beautifully simple. The service drop cable takes a few turns around a ferromagnetic transformer core. That's all! How does it work? If there is no net current on the service, nothing happens. If there is net current on the phases and neutral, its field, concentrated in the core, induces a counter current on the neutral conductor. This has the effect of lowering its impedance to the neutral current generated by the building loads so that when that current has the choice of going back through the water system or the service neutral conductor it is sucked into the neutral by the lowered impedance. You could also say a counter e.m.f is generated at the windings.
At present this simple device, which was being field tested by a New York utility, can be installed on the transformer pole; theoretically it could be used on the customer's (load) side of the electric meter; however the NEC would have to draft an article to allow it. I believe we will be hearing more about this simple solution in the future.

For those who want to better understand the part the earth plays in providing a parallel path for returning neutral current, and how voltages vary widely in the ground at different points of the distribution system I recommend *Ground Currents and the Myth of Stray Voltage* by Seevers.

Magnetic fields from CATV cables

Phone and TV cables have grounding shields which are grounded at the electric pole or transformer. The run of the cable which enters a building and goes to the receiver must have its sheath grounded to the building service point grounding system. If it is grounded to a separate rod, that rod must then be bonded to the building service ground. The reason for this grounding connection is that if there were a nearby lightning strike, if the cable sheath is grounded at a different spot from the building ground, a large voltage can exist during the strike between the cable sheath and the grounded receiver case. This could cause the lightning to "fry" the receiver or TV set and possibly set the house on fire.

But this grounding scenario sets up a parallel path for neutral current to return to the transformer by way of the cable shield to a neighbor's house and hence by way of the bond and the neighbor's service neutral back to the transformer. The situation is electrically the same as the water pipe circuit, though a cable sheath will carry fewer amps than a water pipe.

How can this unplanned neutral current be kept off the cable shield? The cable company doesn't want it, as it can induce hum, as well as being a safety hazard for the cable technicians if the building's neutral is lost.

Here is a suggestion to the CATV technicians: What I am about to describe is not within the jurisdiction of the NEC since it occurs outside the building before the grounding block. Since the incoming cable is already grounded at the transformer or pole, they can insert a "balun" or "isolator" on the street side of the grounding block. This allows the cable frequencies to pass through but essentially blocks 60Hz. One source I found is www.naval.com and the isolator allows 100kHz – 1000 MHz. $10 each. Also mentioned on www.epanorama.net/documents/groundloop/antenna is the use of two RadioShack 75 to 300 ohm antenna transformers, connecting the 300 ohm ends together. The RadioShack #s given are 15-1140 and 15-1253. When testing these types of isolators make sure their grounding conductor is not continuous.

This advice is outside my field and not covered by the NEC. So take it as suggestions that may or may not be permitted by CATV company guidelines. Do not quote these suggestions as authoritative pronouncements. A lightning strike could damage the isolator, but since it is on the outside of the house and easy to replace, perhaps that is acceptable.

Another way to deal with magnetic fields from CATV cables is to re-route the cable away from inhabited areas.

Chapter 8

HARMONICS AND MAGNETIC FIELDS

An issue which has received a lot of attention from electrical trade magazines in recent years has been the harmonic distortion problem. The problem shows itself as overheated circuits due to overheated neutrals, hot conduits due to induction, connector damage from overheating, and overheated and failing transformers.

Harmonics are higher frequencies which are multiples of the fundamental frequency, 60 Hz, which occupies the position of the first harmonic. Thus the third harmonic, by far the most common, is 180 Hz. "Triplens" harmonics, which play a greater role than others in harmonic problems, are found by multiplying odd harmonics by three. Thus 3rd, 9th, 15th, 21st, etc.

The harmonic frequencies which cause these problems are generated by non-linear loads such as computers, fluorescent lights (particularly the new electronic ballasts), most electronic motor controls, various office machines: in short, by the contemporary methods of power control ("switchmode power supplies"). These control devices function by taking "bites" of power out of the smooth 60 Hz sine wave that the utility supplies us with. This injects triplens harmonics back into the system and particularly onto the neutral. The harmonic usually carrying the most current is the third, 180 Hz. A circuit in a commercial building may actually be carrying more current at 180 Hz than at 60 Hz.

The reason for the overheating can be explained in this way: In a three phase line which is carrying more or less balanced loads, the phases cancel each other so that there is very little resultant. This small resultant shows up on the neutral. When three phases are each carrying 10 amps at 60 Hz, there will be 0 amps on the neutral. But when 180 Hz is injected into the system by an electronic non-linear load on each phase, the 180 Hz waves add rather than cancel, since the peaks coincide. Whereas when 60 Hz waves are emitted by three phases, peaks never coincide, and so there is canceling, with 180 Hz emitted by 3 phases the peaks always coincide and you never see a peak simultaneous with a valley.

Whether or not the reader is following the geometry of the wave relationships, the result is clear: if three phases of a circuit were each carrying 10 amps at 180 Hz, the neutral would be carrying 30 amps. In practice you do not have pure 180 Hz when the supply frequency is 60 Hz, but Underwriters Laboratories (UL) tests have shown that a balanced three phase circuit supplying computers may have a neutral carrying 175% of the current carried by each phase. Since neutrals have been previously sized to carry less than each phase, the overheating is explained. Actually, the heating problem is worse, because at 180 Hz the "skin effect" of the higher frequency means that the current can only make use of the outer layer of the copper conductor. This leads to further heating because it is as if the conductor were smaller.

The advice now given in such magazines as EC&M is that the neutral conductor in a three phase system that may supply non-linear loads be sized at 173% of the phase size. Another way of dealing with the problem is to supply one neutral conductor per phase. There are ways of filtering out the 180 Hz and other triplens harmonics from the line but that is beyond the scope of this book.

The question I am leading up to is: do these harmonics cause a magnetic field problem? In other words, if you walk into a building with harmonic problems, will there be elevated milligauss readings on your gaussmeter? The answer is a qualified no, but *if there are net currents* from wiring errors, the magnetic field problem will be greater because the shunted neutral current will be of a greater magnitude.

The magnetic fields from the 180 Hz will cancel out with the 180 Hz from the phases, but if any of that neutral gets loose due to the errors we have outlined, the problem is magnified. EC&M (Sept. 1994) reported on a case where conduits were overheating. They were carrying circuits to fluorescent lights with new electronic ballasts, which generate harmonics. The conduit was measured at 100° F. If I were a consultant on this case I would first check to see if there was net current coming from this circuit, which might account for much of the heating. I would address that problem first and the overloaded neutral second.

The 180 Hz problem emphasizes the necessity of using a gaussmeter that measures the field from 180 Hz correctly. The meters recommended in this book do that, but many others, particularly the less expensive ones usually do not. They either do not measure it at all, measure it too low, or measure it two or three times too high.

When I say that harmonics per se do not cause high magnetic fields, I should mention one exception, and that involves water pipes as grounding electrodes. Since there will always be more neutral current on systems supplying non-linear loads, the portion of the neutral current being shunted to the water system by way of the grounding electrode conductor will also be at higher amperage than if the loads were all linear. This means higher magnetic fields on the water system and also higher net currents on the service drop or service lateral. More reason to insert a dielectric coupling in the water supply pipe.

Chapter 9

DETECTING AND MEASURING MAGNETIC FIELDS

Detection

For simple detection a suitable coil going to an audio amplifier will pick up magnetic fields from net currents. Radio Shack sells a telephone listening coil and amplifier which works for this. Their "Snap-On Amplifier", part # 43-229 is the most compact unit to carry in your pocket. You can put it to your ear as you walk around. A more sensitive version mounted on PVC pipe and using earphones, the **MAGSTICK** tracer unit, is available from MSI (see Instrumentation chapter).

Measurement

A gaussmeter is used. It measures AC magnetic fields in milligauss (mG). Both digital and analog meters can be used. Among the lower cost meters only two-piece single axis meters are practical, as the sensor probe must be free to be oriented while the meter body is held still for reading the display. The meter must have a resolution of one tenth of a mG to be useful. It should also be able to measure the field from the common third harmonic correctly. See the chapter on Instrumentation for a complete discussion.

A more expensive 3 axis gaussmeter is useful for speeding up the initial survey process, though it is no more accurate than a correctly used single axis meter. [Caution: beware cheapie "3 axis" analog meters, currently best sellers but quite inaccurate in many situations].

It is not necessary to orient a true 3-axis meter to the field, as the meter calculates the resultant automatically. On the down side, it is harder to get a directional fix on the source using these meters.

Is it the electrician's job to survey magnetic fields?

You may be called in after an EMF consultant has already measured the building and found elevated fields due to wiring or grounding problems. If you work together on finding the problem you may not need a gaussmeter. But most electricians I have worked with end up purchasing their own meters so they can know when they have fixed the problem, and for trouble shooting other problems.

In the future I believe it is likely that every electrical inspector will have a gaussmeter and electricians may want to know ahead of time if someone has goofed with one of the circuits or if a carpenter or appliance installer has caused a neutral/ground short.

Electricians who work on magnetic field reduction may wonder whether they should go into the magnetic field survey business as a sub-specialty. There is an advantage to being able to fix any problems you find, perhaps on the same visit.

This combination may also be seen to involve a conflict of interest unless your client knows you or your reputation. However many industries combine detection and remediation in the same company.

Using a clamp-on ammeter

Your clamp-on ammeter actually measures the magnetic field from the wire or circuit and is calibrated to display the result in amps. This works because the magnetic field is directly proportional to amps, whether from a single conductor or as net current from two or more conductors. You will be using your ammeter to find net current and to detect miswired circuits. It is useful to have a small clamp-on to get into panel boxes and a larger one to clamp around conduits, water and gas pipes and entrance cables. See **Instrumentation** in Appendix A.

Chapter 10

CONDUCTING A MAGNETIC FIELD SURVEY

You can skip this chapter if a consultant has already done this and called you in to make the electrical corrections.

If you take on this job I have found the following procedure to be useful:

1. Using a gaussmeter on the milligauss scale, take measurements outside the building: the four corners or more will do. You can also take them inside the house with the lights off. This will tell you if there are any external fields present such as from power lines or from net current on street water pipes. These fields are very uniform and weaken slowly as you move away from the line source. Not many measurements are required to characterize the field, since building walls do not affect magnetic fields. Steel building frames such as in commercial buildings may lessen the field from external sources, but the measurement pattern will show you the small variations.

2. Ask for a simple floor plan of the building if available. If not draw boxes to stand for rooms. Have all the lights turned on in the building - not just one section or one floor at a time. A first floor circuit running in the ceiling may affect the second floor living space. To check out the receptacle circuits either turn on some floor lamps or carry a small heating coil to plug in to a receptacle when you measure in a room.

Take measurements at the four corners of each room, the middle, and perhaps a few more if there is a field present. (If you get into computer imaging, measurements in a grid every 6' in both directions should be adequate. See Chapter 12). The convention is to take them at waist level about 3' from the floor. Keep about a foot away from the wall when taking side-of-room measurements. If you measure an unusual field you can see if it is coming from floor, ceiling or wall. Write the mG numbers down on the floor plan. A legal-sized clipboard with Velcro at the bottom for your meter will make this easier. A 3-axis meter such as the Bell 4080 makes this part of the survey go very fast.

When the survey is complete you will see a pattern of elevated fields if a circuit or pipe carrying net current is involved.

The survey is useful not only as an aid to locate problem sources but as documentation of the existing situation. You can do an "after" survey and compare the two. See **Figure 12** for before-after surveys of an elementary school with an accidental neutral/ground short.

3. The next step is to trace the path of any field source you have identified. Use a single-axis meter to get a fix on the line source. This becomes easy as you get used to the fact that the line source will be at right angles to the long side of the probe and one flat side will be looking at the line.

A quicker way to trace a line source is to use the MAGSTICK tracer unit. The path of the circuit or pipe is easily followed as you pass the unit over the line source and listen to the loud hum with the headphones. You can also get a precise directional fix at any point, which is useful where a circuit changes direction. Using the MAGSTICK speeds up the tracing work and can locate deeply buried conductors as long as there is a net current on them. **Figure 13** gives an idea of how the MAGSTICK is used.

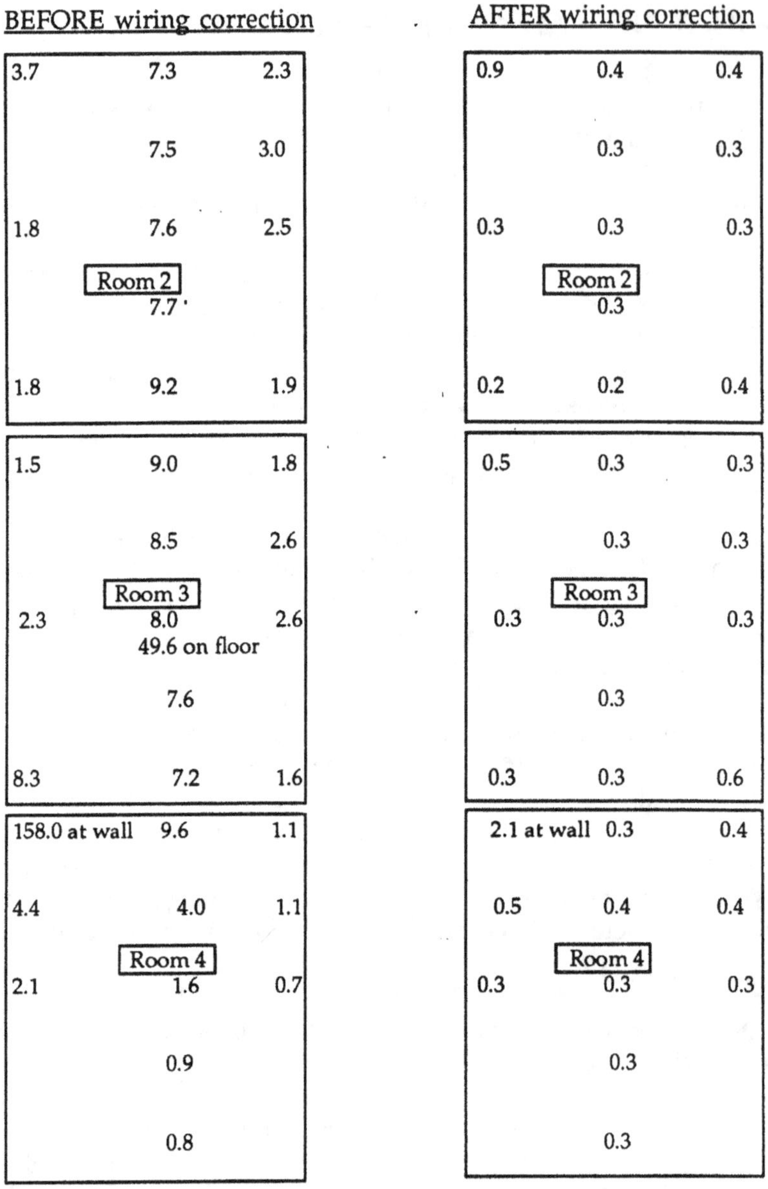

BEFORE wiring correction   AFTER wiring correction

3.7	7.3	2.3
	7.5	3.0
1.8	7.6	2.5
	Room 2	
	7.7	
1.8	9.2	1.9

1.5	9.0	1.8
	8.5	2.6
	Room 3	
2.3	8.0	2.6
	49.6 on floor	
	7.6	
8.3	7.2	1.6

158.0 at wall	9.6	1.1
4.4	4.0	1.1
	Room 4	
2.1	1.6	0.7
	0.9	
	0.8	

0.9	0.4	0.4
	0.3	0.3
0.3	0.3	0.3
	Room 2	
	0.3	
0.2	0.2	0.4

0.5	0.3	0.3
	0.3	0.3
	Room 3	
0.3	0.3	0.3
	0.3	
0.3	0.3	0.6

2.1 at wall	0.3	0.4
0.5	0.4	0.4
	Room 4	
0.3	0.3	0.3
	0.3	
	0.3	

Figure 12. A kindergarten and two elementary classrooms showing the milligauss measurements at 3' above floor level from a net current in the circuit under the floor due to a neutral/ground short in room 4 . Before and after correction. Floor levels were between 40 & 50 mG.

4. After tracing the problem circuit(s), take time out to think. A coffee or lunch break often saves time in the long run as ideas come together. Does the line source run through the basement? Does it seem to be where a water pipe might run, or is it clearly an electrical circuit, running to receptacles or switches? Are there two parallel line sources going up the wall above a light switch? If so, suspect knob-and-tube wiring.

Take a look in attic crawl spaces or in the basement for visible clues. Is there a magnetic field on the water pipe where it enters the building? On the gas line?

Don't spend too much time following these clues, since you will be able to pin down the sources as soon as you go to the service breaker panel or subpanel. But if something is there to be seen, pick up on it.

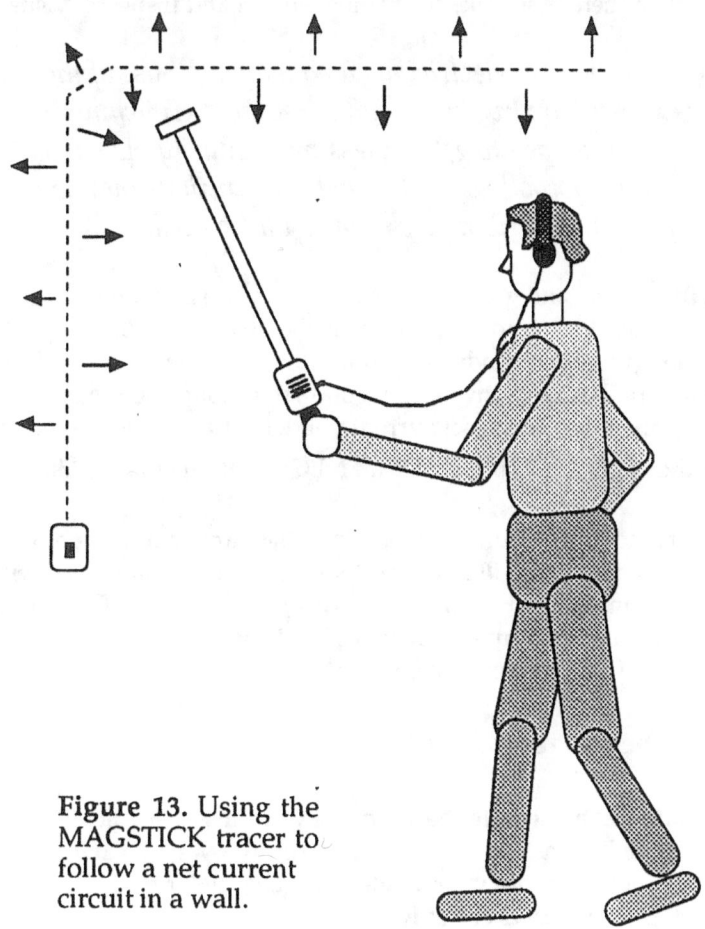

Figure 13. Using the MAGSTICK tracer to follow a net current circuit in a wall.

The MAGSTICK tracer is without doubt the most useful tool for tracing net current in any conductor. It does not need a signal injector because the net current provides the signal. Listening to the frequency signature of the field tells you a lot about its source. It can trace water pipes underground and follow the water mains in the street. One 9V battery powers the unit.

Chapter 11

DIAGNOSING FROM BREAKER BOXES

Did you trace a net current circuit to a sub panel or to the service entrance breaker box? If it was to a sub panel go there first. Take the front cover off and inspect it visually.

CAUTION: *For the non-electricians who may make use of this book, removing a breaker panel cover exposes bare live buses and connectors. Touching them could result in shock or electrocution. It is easy to get careless momentarily. Let an electrician do this part. Electricians: you must follow all the safety precautions that your company and OSHA require. I am simply describing techniques I have used.*

First check the neutral and grounding buses if they are there. If this is a subpanel, is there any evidence of a neutral/ground connection? Is the neutral bus bonding screw turned in? Are there any equipment grounding conductors connected to the neutral bus? Is there a bonding strap or wire to the neutral bus? Are any neutral conductors connected to the grounding bus? Any of these could be the cause of the net current you have traced and can be corrected at the panel.

Grounding the neutral at a subpanel is an NEC violation (Article 250-24(A)(5)).

If you suspect a neutral/ground connection at the panel but it is too overcrowded to see clearly, there is an instrument that will indicate a neutral-ground connection within about 10 conductor-feet of where the instrument is connected. This is the SureTest Circuit Analyzer or the SureTest Pro (see Instrumentation chapter).They need to be used with alligator clip adapters, since they are designed to be plugged into receptacles.

Identify the net current circuits

If you have followed a circuit to the subpanel you can run a quick check to see which circuit(s) has net current. Either pass your MAGSTICK sensor over the cables entering the box and listen for the loudest sounds, or use your single-axis gaussmeter probe and watch for a high reading. This is a quick preliminary check only.

To verify and measure the circuits carrying net current, use your clamp-on ammeter. Start with the cables you suspect. If it is easier, clamp around the whole cable or conduit outside the box. Otherwise you have to work your clamp-on around the group of wires in the circuit just inside the box.

Remember, <u>a correctly wired circuit will read zero on the ammeter</u>. Anything above 5 mA net current would trip a GFCI and indicates a wiring problem. If the box is crowded and you can't clamp around both neutrals and hots at once, alternately clamp around the hots together and record the load; then clamp around the neutrals. If there is a difference, there is a net current. *This is one of the most important paragraphs in this book. Please re-read if necessary.*

If loads are fluctuating as you measure, you will get a better comparison if you use an averaging ammeter, such as the Fluke 33, or connect the clamp-on probe to a meter which has max/min/av in the record mode.

Even a half-amp net current will create some possibly objectionable magnetic field, particularly within 3' of the conductors, but the usual net currents we see are in the 1.5 to 10 amp range, with

3 amps being common in a miswired residence. Commercial buildings may have higher net currents. A 3 amp net current produces 6 mG at 39", 3 mG at 78". Measuring directly on the conductor with a gaussmeter you may see up to 300 mG, which is roughly equivalent to 3 amps on the conductor.

Once you have located a circuit with net current, record the amperage (let's say 3.4A). Then go looking for its net current buddy - the circuit carrying the missing (or extra) 3.4A.

If this is a case of GCs from different branch circuits connected together (we call them "parallel neutrals") you should be able to find another circuit with about 3.4A net current on it. Net current can be caused by either a deficit or an excess of neutral current. The reading will be the same. If you do find a pair of branch circuits with approximately the same net current you can compare the hots and GCs of one circuit and then do the same with the other circuit. In our example, one will show the GC is carrying 3.4A less than the hot(s) whereas the other circuit will show the GC is carrying 3.4A more than the hot(s).

If you have found a pair of circuits with the same net current, you have the problem diagnosed, and now all that is left is to locate the junction box where the GCs are connected.

There is an alternate way to locate paralleled GCs which you may prefer. In some cases the circuits may be hard to clamp on to, or there may be no load on them at the time. Use a signal injector and tracer (see Instrumentation chapter). Trip the breaker or unscrew the fuse on the circuit you have found with net current. Disconnect the GC conductor (see CAUTION).

CAUTION: *If you do have paralleled neutrals, even though you have tripped the breaker for this circuit, the neutral may be carrying current from the other circuit which may have a load on it and so will be at 120V to ground when disconnected. The only safe way to do this is to trip either the main or the other breakers on circuits it may be connected to. Only a professional should disconnect a neutral.*

If you have determined that it is safe to disconnect the GC, and there is no load on the GC, inject a signal onto the disconnected GC. With the tracing unit find the other paralleled GC in the panel box. If the signal does not return on another GC try the equipment grounding conductors to see if a neutral/ground short is indicated.

As time goes on I find less need for signal injection since tracing energized circuits is usually successful. However at the times they are needed they are invauable.

In commercial buildings and schools it is possible (though infrequent) that a paralleled GC will return to another subpanel. The magnetic fields in the offices were very high (see **Figure 15**).

For a clear animated visual presentation of how net currents are created and how to trace the errors, see my video, **Tracing Magnetic Fields in Building Wiring.** See **Videos** in Appendix.

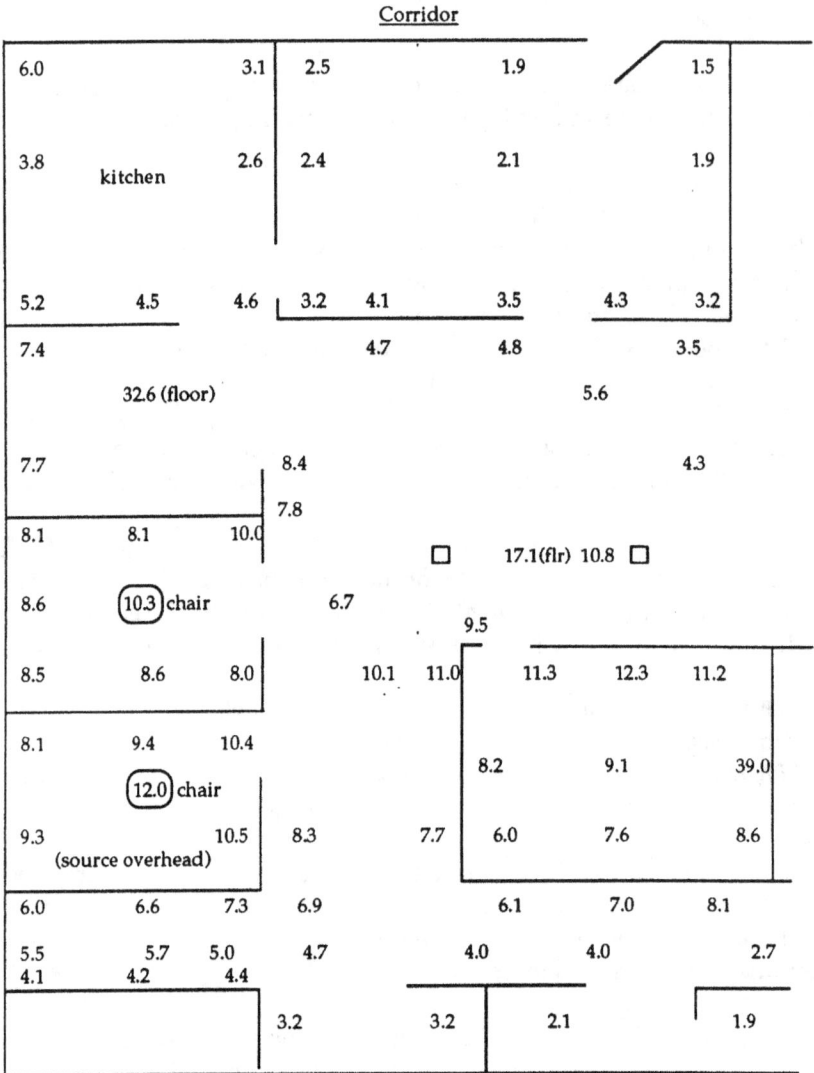

Figure 15. Milligauss levels in an office suite due to multiple paralleled neutral connections

If there is no paired net current circuit

Suppose you can't find another circuit with that 3.4A net current on it. This would indicate either that you have a paralleled GC returning to another panel or else you have a GC/ground connection out there where 3.4A is being shunted to grounding paths. **Figure 15** shows the magnetic field levels measured in an office suite due mainly to paralleled neutrals, some of which went to different subpanels. See **Figure 19** for a computer-generated contour map of these fields.

When GC is shorted to "ground" out in a circuit, most of it may transfer to a pipe, duct, steel beam, etc. so that very little returns on the equipment ground.

Another way to verify a GC/ground connection is to use an ohmmeter and measure the resistance from the disconnected neutral to the grounded box or bus. It should be infinite if there were no short to ground or a grounded neutral. A short will show as a low value on the meter. If you know your meter and the resistance per foot of the GC (see your NEC tables) the ohm value may give you an idea of how far out into the circuit the short may be. For instance, the AC resistance and reactance for a #14 conductor is 2.7 ohms per 1000 feet, or 0.27 ohms per 100 feet. A #12 conductor is 1.7 ohms per 1000 feet. It may be easier to put a signal on the GC and start tracing.

If no one else is in the building at the time, an electrician may want to verify the GC/ground short in another way. I do not recommend this! There is a potential shock hazard for someone in the building. I will give an example. I was called to a California school to find the source of an unusual magnetic field which ran down the center of a classroom. Working with the school electrician I identified the circuit with the net current that was causing the field. There were 11 amps missing from the neutral. Because he knew that no one was in that section of the school at the time (after hours) the electrician disconnected the GC but left the circuit energized. The lights stayed on. I found 2.5 amps of the missing 11 amps on the insulated equipment grounding conductor returning with the circuit. As we found after tracing the circuit with a MAGSTICK, the rest of the shunted GC was returning to the main through building pipes and steel posts. The N/G short was at a junction box in the ceiling. See **Figures 12 and 18** for floor plans of these classrooms.

There is another indication of a N/G short in a circuit originating at a subpanel. Clamp an ammeter around the feed circuit supplying the panel. There will be a net current due to shunting of some of the GC current to pipes, etc. which are not returning the current to the subpanel but directly to the main panel.

Verifying neutral current on the water service pipe

Neutral current on the water supply pipe can be measured with a clamp-on ammeter with large enough jaws (see Instrumentation). Otherwise it will show up on a gaussmeter held against the pipe. The sensor from an MSI gaussmeter can give an indication of the amperage on the pipe. There is an approximate 100-1 ratio. 100 mG indicates about a 1 amp current.

A similar if not identical net current will show up if you clamp your ammeter around the service entrance cables. If you can't get your ammeter clamp around the whole cable, you can clamp around the phases in the box and compare that reading with what the neutral is carrying. If the phase imbalance is 12 amps and the neutral measures 8 amps, there are 4 amps of neutral missing. This will usually be found as a 4 amp current on the water supply pipe. You can start by looking for it on the grounding electrode conductor leading from the service entrance to the clamp around the water pipe.

If the current is present on the grounding electrode conductor but not on the water pipe, which might be the case if the water supply pipes have been changed to plastic, trace the grounding electrode conductor to find other grounding connections. They might be to a gas pipe, telephone cable grounding sheath, cable TV ground, or anything else which allows the neutral current to find a metallic pathway to return to the transformer supplying the building. You will seldom measure any significant current going into earth through a ground rod. The resistance of the rod/earth interface is usually too high to compete with the metallic conductors carrying current back to the transformer.

If there are amps on phone cables or TV cables you can call the company involved and their technicians will usually be cooperative about helping to get the power company's neutral off their line. CATV companies use a neutral blocker which stops low voltage current but will allow voltages from lightning to flow.

If the neutral current is on the gas supply pipe then the gas company should be willing to install a dielectric coupling at the gas meter. Many gas companies have already done this, as they don't appreciate the spark hazard on their lines.

Before 1990 gas pipes in the ground were listed as qualifying grounding electrodes. The 1990 NEC disallowed them as grounding electrodes but still required the inside pipe to be bonded to the neutral/ground point in the building for internal safety reasons. This left it up to the gas company to put in insulating (dielectric) couplings outside the building at the gas meter so as to keep any fault current or stray neutral current from traveling through the gas supply lines.

If there is current on the water pipes the recommendation made previously can be carried out, with the following caution: the reason there is substantial current on the water pipe may not only be the low impedance of the water system, but a high impedance at the service neutral connections, either at the service drop at the building or at the power pole or both. So *before* the plumber inserts a dielectric coupling in the water line (remember, 10' or more outside the building) ask the owner to call the utility and ask them to check their neutral connections. Usually they will do this willingly, particularly if you mention "voltage problem". Many utilities have no maintenance schedules or inspection routines for service drops and rely on customer complaints to let them know when there are problems.

There is a quick way to predict whether there may be a bad connection in the utility's neutral. Clamp your ammeter around the service neutral. Measure amps. It is better to allow a few minutes of measuring time using a multimeter which records an average value, since neutral current fluctuates. Now clamp around the grounding electrode conductor going to the water pipe. Compare the two amperages. This will tell you the relative impedances of the two return paths.

If the amps on the neutral are, say 12A, and on the water pipe 2A, then there is not much reason to believe there is anything wrong with the neutral connections.

But if the water pipe is carrying, say, 4A and the neutral 8A, I would suspect corrosion at the neutral connections. This is not an exact test but it gives you some indication of neutral connection problems.

I have seen service drops where the neutral "connection" at the line was just a loose twist of the cable. Nowadays they use connectors. The reason for having the utility check and clean their neutral connections before a plumber inserts a dielectric coupling is that if the utility's connections are poor you may have voltage problems after the plumber inserts the dielectric. "Voltage problems" means one hot leg will show higher voltage than 120V and the other leg will show lower voltage. The plumber also may be at risk when he cuts through the pipe to insert a spacer if the utility's connections are poor. Some plumbers clamp automobile jumper cables across the pipe where they are going to cut it to protect themselves from any voltage that may be present. You might suggest this to the plumber.

Sometimes when the utility linesman cleans and re-clamps the service neutrals the water pipe current will all but disappear. I recommend that the dielectric coupling still be inserted to protect the building from future changes which could reestablish a neutral current path through the water system.

Sometimes there is *excess* neutral on the service drop. In this case a neighbor's neutral current is coming into the client's house through the water pipes and using the better neutral connection of the client's service drop. In this case, having the neutral connections cleaned may invite more of the neighbor's neutral into the house, which emphasizes the need for the dielectric coupling in the water pipe.

One way to verify this situation is to shut off the main disconnect to the building (after being sure that no one is using a computer, etc. and having obtained permission from all the occupants). Now measure the current in the service neutral and also in the grounding electrode conductor. If there is current, it is coming into the building through the water system and continuing out through the service neutral, which is never switched.

Eventually the utilities may deal with neutral connection corrosion as a regional problem and check all neutral connections on their lines as part of a maintenance schedule.

Chapter 12

LOCATING THE MISCONNECTION

The paralleled neutral

Once you have identified the two circuits which have their neutrals connected somewhere in the circuit, you need to find the spot and reconnect them correctly by separating the circuits (two wire nuts instead of one). If you did the wiring yourself perhaps you can remember where these two circuits share a junction box, switch box, etc. Any existing wiring diagrams may help pinpoint the possible boxes. If someone else did the wiring you will need to trace the path of the GCs.

There are two ways to proceed:

1. Leave a load on one of the two circuits and trace the path of the net current with your MAGSTICK or single axis gaussmeter.

2. Switch both circuits off, disconnect one GC and inject a signal on it. Follow the signal with the tracer unit.

Using either method, trace the net current or tracer signal out to the farthest point away from the breaker box. Look for a junction box that is in a position to be shared by both circuits. For instance, if one circuit goes out along one side of a building and the other serves the other side, look for boxes at the end of both runs where they may come together. Perhaps the box is used to connect an additional circuit to outside lights, etc.

An excellent electrician I worked with on a large school campus preferred to start tracing at the panel. Since the problem circuits were usually lighting circuits in conduit, he went from junction box to junction box in the raised ceiling. He verified that he was tracing the problem circuit by clamping his ammeter around the conduit and seeing net current. He then visually inspected the junction box connections. Of course it was usually the boxes with four or more conduits that were to be suspected.

Use the ammeter to clamp around each conduit to see which have net current. Then open the box and look to see how the GCs have been connected.

The box you are looking for may be behind a multiple light switch; it may be a junction box in a raised ceiling or attic, or perhaps in the basement or crawl space. When you open up the box you will see that all the GCs have been twisted together with a large wire nut. (See **Figure 14**).

The correction is to separate the GCs according to circuit. If there is a load on one of the circuits, separating GCs can expose the electrician to 120V between the load GC and the GC returning to the breaker panel. Some electricians are used to working with live wires and have developed techniques that protect them (usually) against getting shocked. Do-it-yourselfers have not developed these techniques and controled movements and should shut off both circuits before separating the GCs. I do not advise non-electricians to make these corrections, but because in many states some classes of non-licensed personnel such as home owners and maintenance personnel are permitted to make wiring changes it cannot be prevented. But they should realize that by holding a GC from an energized lighting circuit in one hand and touching either the return GC or the grounded box with the other hand, a lethal current may pass across the chest through the heart and cause the heart to stop. A GC is at close to zero potential only when connected to the service panel neutral bus, which in turn is connected to the service entrance neutral.

Be prepared for surprises when you open up a junction box. Other errors may appear. It is possible that a circuit has been added, and the electrician had used a hot conductor from one branch circuit and a GC from another. This will cause a large magnetic field from the consequent loop. Another odd connection that causes a net current magnetic field is the connection of an extra GC from another circuit which has been discontinued; that is, the electrician disconnected the hot conductor and capped it, but did not bother to disconnect the GC, and so the return current on the live circuit travels back to the panel on both GCs. The extra GC should be disconnected and capped also.

Sometimes the error has been made in a duplex receptacle which is fed by two circuits: one circuit feeds the half of the receptacle which is always on and the other circuit feeds the switched half of the receptacle. If the tabs on the sides of the receptacle have not been broken off as they should be the neutral tab will connect the GCs from both circuits which will then be in parallel. See Case Studies chapter for one example, and **Figure 8** for a picture of the receptacle and the location of the tab.

As soon as the GCs have been correctly connected and the circuits energized you will find that the magnetic field you had been tracing is gone, which means that the magnetic field from that wiring is no longer detectable (less than 0.1 mG) a couple of feet from the wires. This is a dramatic change the first time you see it. First you are detecting magnetic fields all through a section of the building; next you are detecting either nothing or a couple of tenths of a mG from overhead fluorescent lights.

If there is still a generalized magnetic field there look for a second source. Sometimes one problem masks another and you have to solve one to get a good look at the other. It can be frustrating to solve a problem and find that there is still a magnetic field there, even if reduced. But the upside is that the second problem is more easily traced with the first problem out of the way. Just continue with the tracing strategy. I am reminded of a Malibu estate with six wiring errors in the main house. Some of these circuits overlapped, so as we corrected one the others were easier to trace. It was a frustrating day at times, but the final result was a building completely clear of area magnetic fields.

By separating GCs from different branch circuits you have corrected two violations: NEC 300-3(b) and 310-4.

The neutral/ground short

The same methods can be used for tracing a N/G short though now you are dealing with only one circuit.

A second method is to turn the circuit breaker off, disconnect the GC from the neutral bus and inject a signal on it, with the other lead from the injector clamped onto the box or grounding bus. The signal can be traced out into the circuit.

Tracing a N/G to the point where the short occurs can be difficult since some of these shorts are accidents and therefore illogical. But you can follow the signal or magnetic field out into the circuit, taking care to notice when the signal starts to show up on metallic paths such as hung ceiling frames, pipes, building steel, sprinkler pipes, etc. Look for junction boxes or switch boxes which may reveal an equipment grounding wire connected to a GC, or else a break in the GC insulation as it comes into the box. Sometimes the cover has been screwed down pinching a GC and cracking the insulation.

Remember that once you have traced out past the location of the N/G short you will no longer see a net current or strong signal on the circuit. One way to narrow down the location is to pick a box that is beyond the suspected grounding point. Open it up. Turn off the breaker for the circuit. Detach the signal injector from the neutral in the breaker box but leave the neutral disconnected. Bring the signal injector to the junction box, detach the neutrals and inject a signal on the neutral which is headed back to the breaker panel. The signal will follow the grounding path(s) rather than go back to the breaker panel in the circuit.

Perhaps the most difficult N/G short to find is caused by a nail or screw that nicks a GC in romex and connects it to metal wall lathe. Because the current spreads out widely through the lathe, when tracing the net current or signal it seems to more or less disappear at the point where the nail is. But this is a helpful symptom if you recognize it.

Another scenario occurs when a nail or screw has gone through a hot conductor shorting it to the ground wire. This is a ground fault and the breaker trips. The electrician tests for ground fault on the conductor at the box but does not want to spend the time to locate the fault or replace the circuit conductors. He decides to switch the white and black conductors, using the white for a hot and the shorted black for the GC. Now the undamaged white conductor is carrying the hot current and the shorted black conductor is carrying the GC current and dumping some of it to a grounding path by way of the nail. I have seen two cases of this. It's a serious Code violation but someone didn't care. It sort of stands out when you open a breaker panel and see a white wire going to a breaker and a black wire going to the neutral bus.

Here are two examples of N/G shorts. In a newly wired multi-million dollar residence which showed high magnetic fields throughout much of the building we found two N/G shorts as well as three paralleled GC circuits. In one case I traced the net current magnetic field to the end of a lighting circuit where I picked up currents on a conduit which touched a gas pipe. The gas pipe carried the amperage around the building, shunting some of it to either grounding paths in two places but delivering half of it back to the main breaker box. This current path created magnetic fields in a child's bedroom before connecting up with the neutral at the bonding point.

And what was causing all this? A neutral/ground short in a wrought iron chandelier installed by lighting specialists. This puts an extra responsibility on the electrician to check the work of the installers. This is also important when appliance men install electric dryers. (See Appendix B for an instance of net current from a dryer room in a public housing building).

The second N/G short was a tough one because it had been covered up. I traced the net current field to a light switch by the sink in a child's bathroom. Using a tracer unit on the disconnected GC, an electrician followed the signal from the wall switch box to a spot on the wall tiles at the sink where normally a GFCI receptacle would be located but which was not there. Evidently it had been tiled over. Since we didn't want to tear out the expensive tiles the remaining light circuit which had been wired through that covered-over box was wired an alternate way, and the conductors to the box disconnected. This removed the N/G short which we assumed was due to an uncapped unused GC in the box touching the grounded box.

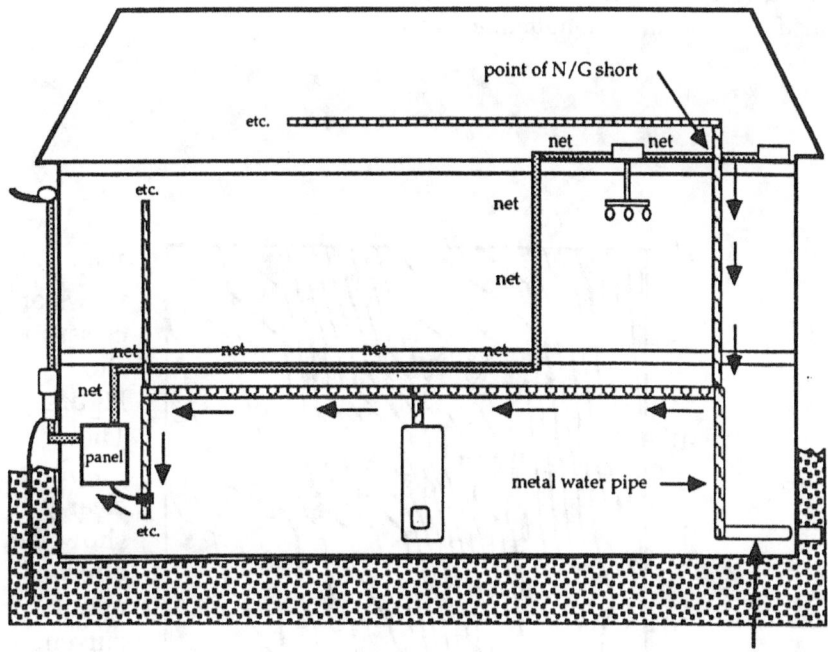

point of N/G short

etc.

net   net

net

net

net

net net net net

net

net

panel

etc.

metal water pipe

Plastic water supply

Figure 16. A neutral/ground short from romex to a water pipe in the attic creates a loop of net current in the house. Neutral is shunted to the water pipe which returns it to the grounding electrode conductor, hence to the entrance panel where it gets back on the service neutral. There is an equal net current in the romex circuit affecting another part of the house. In this case the house is supplied by a plastic water system and so no neutral leaves the house on the water pipe. The problem is entirely internal.

Using your head

It is easy to get caught up in tracing circuits, particularly complex ones, in a state of fascinated confusion. Sometimes you will be dealing with two errors partially overlapping each other. I have found it important to take time out to think the situation through logically. As I mentioned earlier, many problems are resolved during coffee breaks or lunch breaks, or while driving home. Logic does most of the work and good hunches and sometimes luck do the rest. Looking back on each job I usually see that logic could have done more of the work.

This is detective work and takes concentration. Some will find that they can trace the error better by working alone and silently. Talking sometimes distracts, particularly when your partner may not understand the process you are following. On the other hand others work better with constant discussion.

If you use a computer and have a good graphics program installed, a useful aid to diagnosis is available. The magnetic field measurements recorded on the floor plan can be entered into the graphics data sheet in columns. If the graphics program has a contour map selection it will graph the data as either a line contour plan or a shaded plan.

Three examples from the DeltaGraph program are shown in the following figures. **Figure 17** shows the high school classroom pictured in **Figure 4**, with a line source under the floor. **Figure 18** shows the three classrooms from **Figure 11**, and **Figure 19** shows the complex situation in the office suite (**Figure 15**) resulting from paralleled GCs. The "X" in the figure shows the position of the junction box with wire-nutted neutrals.

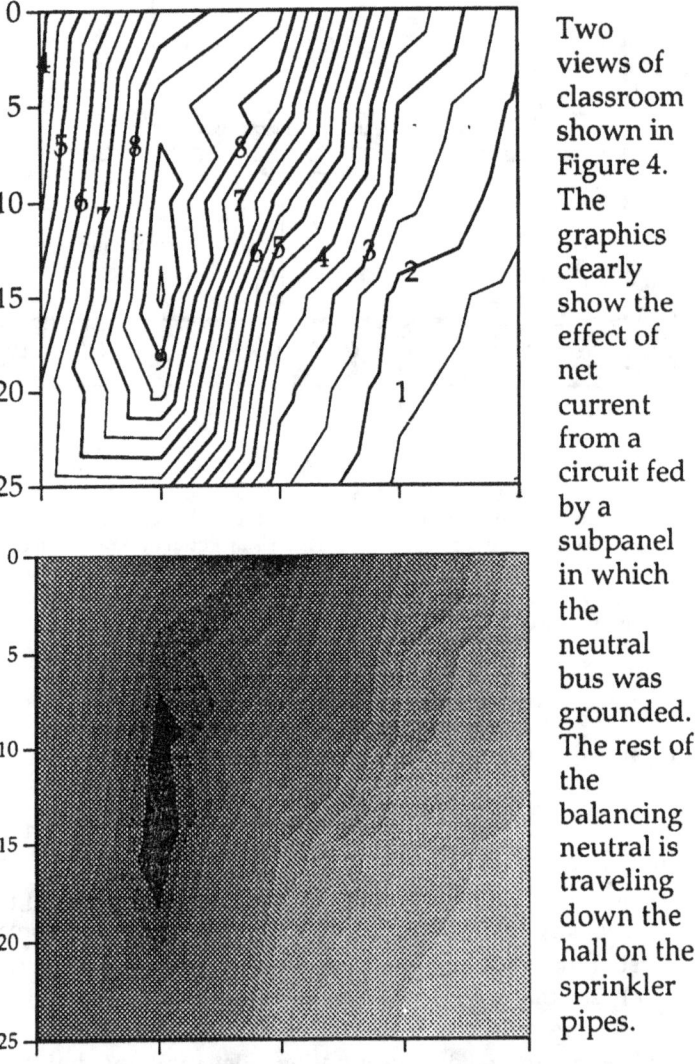

Two views of classroom shown in Figure 4. The graphics clearly show the effect of net current from a circuit fed by a subpanel in which the neutral bus was grounded. The rest of the balancing neutral is traveling down the hall on the sprinkler pipes.

Figure 17. The high school classroom fields before correction. See also Figure 4.

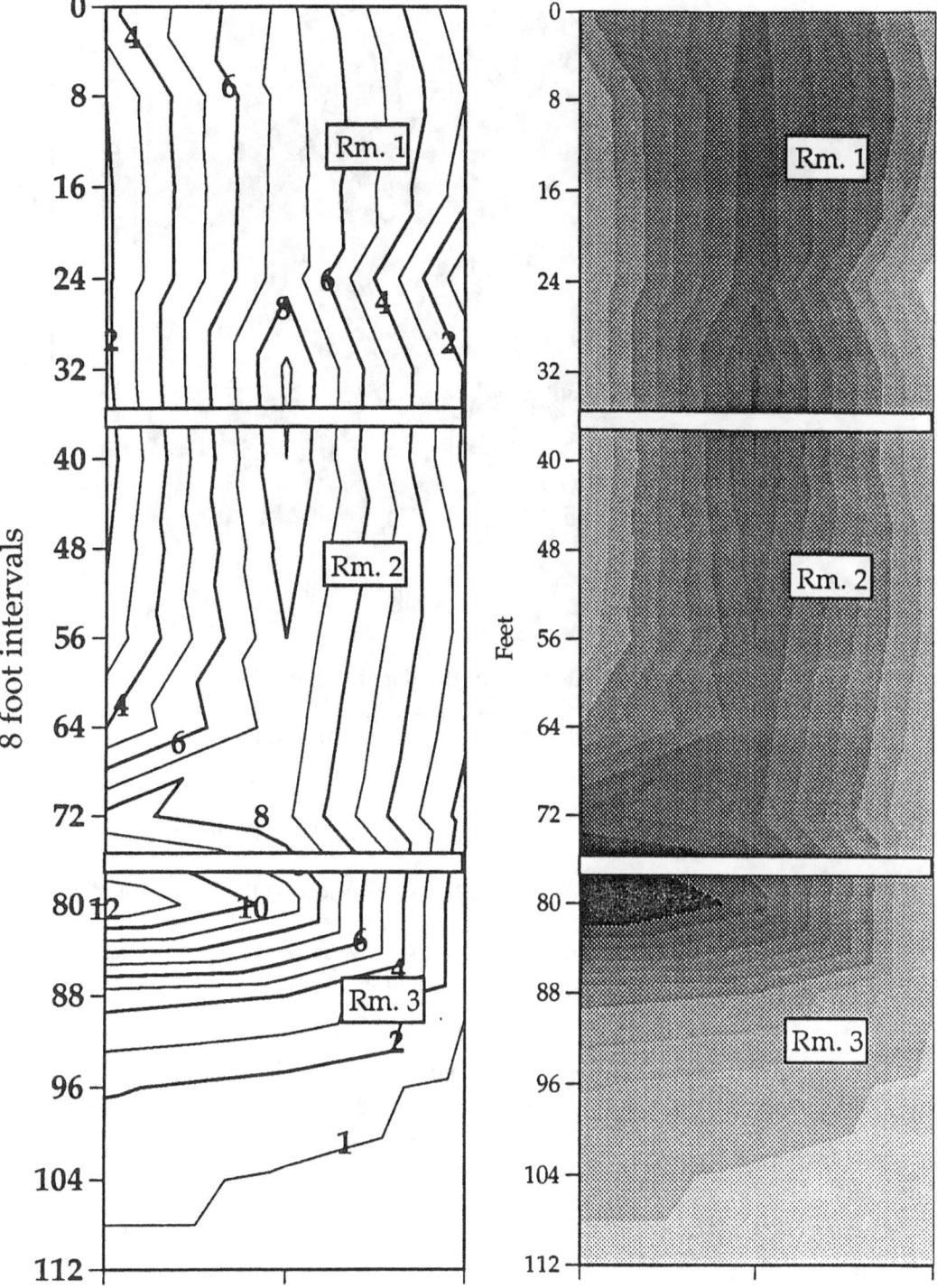

Figure 18. The kindergarten/elementary classrooms pictured in Figure 12 before the N/G short was located and corrected.

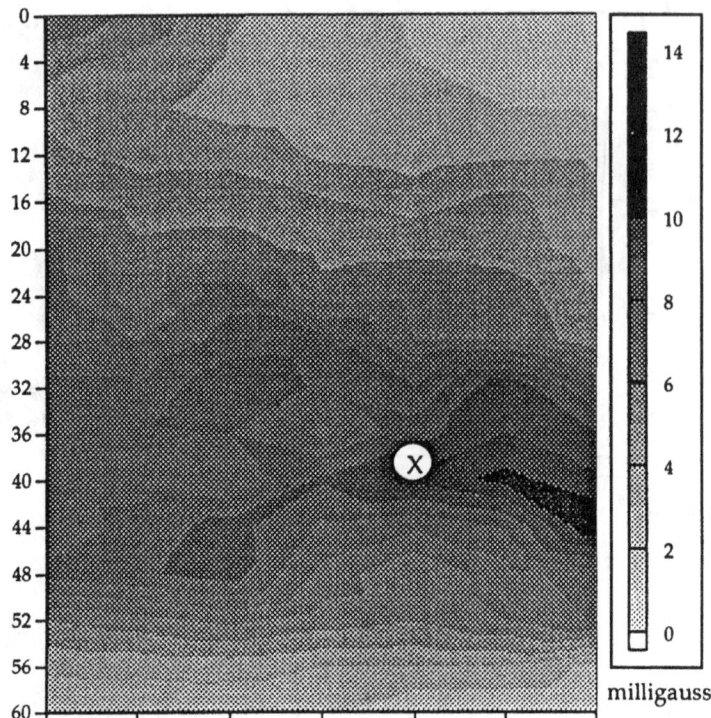

A alternate view of the data in Figure 15. Gaps in data
were filled in to achieve graph. Measurenents should be
taken without gaps when graphing is intended.

Figure 19. The office suite affected by paralleled GCs. The "X" shows one major error site.

Do these graphics have diagnostic value? Having produced them at a later date I can only say
that I can imagine them speeding up the diagnostic process if they had been available at the site,
particularly in complex situations. To have them available at the site is a feasible option if one has
a laptop PC with an adequate graphics program. Entering the data does not take long, and the
graphing is instantaneous.

As a visualization tool for presenting the situation to the client this graphics tool may have its
greatest use. A before-after series will be very impressive if you were successful.

Chapter 13

SHIELDING AND FIELD REDUCTION IN HIGH CURRENT CONDUCTORS

You may be asked if a room can be shielded from some source of magnetic fields. Shielding is a useful last resort and has its place in solving problems. First correct any errors or incorrect grounding situations that exist. Then deal with whatever sources are unavoidable. It should be understood that passive shielding with materials has no effect on net current fields.

Sometimes small or medium sized transformers for low voltage systems such as alarm systems have been located where they are just behind a wall or just under a floor where children or adults spend time. Since they give off high fields locally you can relocate them by a few feet and solve the problem.

Small analog clocks such as are in the front of kitchen ranges give off high fields, often higher than the electric stove burners. You may be asked to disconnect them if not necessary for the cooking process.

One situation exists which may call for shielding: when a breaker box is located next to a bed or other occupied spot, and relocating the bed or desk is not feasible. Electric panel boxes give off locally high fields because the conductors are separated inside. Boxes can be wired in such a way as to keep conductors of all circuits together until they must diverge. Some load centers now on the market have neutral buses running parallel with the breakers so that the neutral is connected close to where the hot is connected (See **Figure 20**). Also, when you bring the service cables into the box you can sometimes keep the phases and neutral together in one loop instead of splitting them into loops on opposite sides of the box. Keeping conductors together reduces the magnetic field from the box by a significant amount.

If the field from the box is still stronger than the client wants, shielding can reduce it. The metal used most often is an alloy with high "magnetic permeability" symbolized by μ, hence "Mu metal", a trade name. There are many distributors of magnetic shielding alloys in various thicknesses from heavy sheets to thin foils. The material works by concentrating magnetic lines of force inside the metal that would otherwise be spread out into space.

If shielding is needed there are a number of companies which specialize in shielding. Some can also supply material in sheets. For shielding companies see Appendix.

Clients may ask you about some device they have heard about that "neutralizes" or "realigns" magnetic fields and produces a "healthy energy environment". The advertising copy for these devices typically speaks of harmonizing the fields, or introducing coherence (or incoherence) in the "photons" of the magnetic field, etc. The explanations may sound scientific to a layman. Studies are cited backing their claims, but never is a reference given as to where the study was published. In general the advertising copy gives the clear impression that the device really does something to the electromagnetic fields. I have not yet tested one of these devices which produced a measurable change in the field or showed any visible change of the wave form of the magnetic field on an oscilloscope.

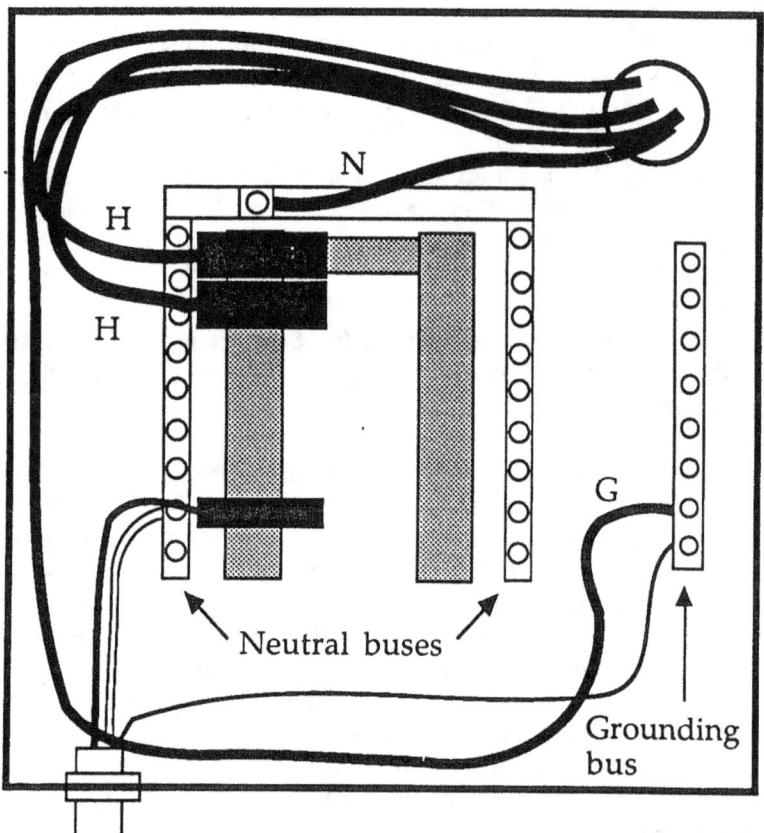

Figure 20. Diagram of a subpanel breaker box which allows conductors to remain together until necessary to diverge. True of supply feed as well as branch circuits. Minimizes magnetic fields.

Massive power

In high rise office buildings another range of problems exists. Electrical distribution rooms and closets with bussways (always a strong magnetic field source) and transformers may be adjacent to office space. Either massive shielding of the room or change in use of the adjacent space may be called for. It may also be cost effective to look into the availability of alternate buses which are more compact. Halving the distance between bus conductors should cut the magnetic field in half.

When transformer rooms and switching cabinets in large office buildings are involved it is best to leave the shielding job to companies which specialize in massive shielding installations.

High current feeder circuits may run in conduits near occupied space. Even though free of net current some of these conduits, trays or busways may generate significant fields. These can be reduced by compacting the conductors in trays or using twisted cable in ferrous conduits. Twisting conductors as a magnetic field reduction technique works only with correct wiring generating no net current. Net current is completely unaffected by twisting. Twisting should be employed using UL approved twisted cable. Ferrous conduits also have a partial shielding effect on balanced circuits depending on how they are grounded. They will not shield net currents as normally installed.

Shielding by induced counter-currents

Since shielding by way of induced counter-currents is potentially a very useful strategy, let's look at how this effect shows up in buildings affected by strong magnetic fields. We are speaking of a partial field cancelation from the counter-current induced on a conductor running parallel to the source conductor.

Let's take an example. If you are measuring in a room with metal window frames and power lines running nearby parallel with the window wall, move your gaussmeter sensor along the wall towards the window frame. As you reach the frame the field strength will go up, but as you move into the space within the frame you will see the field go down relative to the level away from the window. There is a small shielding effect within the window space.

What is happening? The AC power line magnetic field lines of force are cutting through the aluminum or steel window frame and inducing a loop of current in the frame which runs opposite to the current in the lines, thus generating an opposing magnetic field which partially cancels the power line field within the current loop produced by the window frame. This effect is the basis for many interesting and potentially economical ideas for shielding, some of which are being proposed by groups like EPRI for sections of power lines.

This effect also accounts for unexpected reductions of field strength within some steel-framed buildings as well as in increase in the field near the steel. At an EMF health effects conference I attended a report was presented describing the magnetic field levels found in a Canadian school district. An overhead was displayed of a field profile of a power line. It traced the field strength across the yard and through the school. The profile showed a distinct dip as it entered the school and rose again at the far edge of the building. The reaction of the EMF professionals in the audience was that there must be some error in the measurement protocol since they would not expect to see a shielding effect from a building structure.

However I have seen this kind of field reduction in a commercial building near a transmission line. When two profiles were taken; one outside the building alongside, and one through the building, the outside profile weakened in a smooth way, but the inside profile showed a dip inside with the minimum level near the center of the building and a secondary peak at the far edge. I investigated and found that a current was running around the periphery of the building from a location in the building footing, which I assumed to be steel rebar or else foundation girders. This current was induced from the transmission lines. It was not ground current. The secondary peak along the rear wall was due to the fact that at the far side of the loop the current was flowing in the same direction as the power line field and so amplifying the field, whereas the current flowing in the near side of the building opposed the power line field and showed a modest canceling effect. The field inside the building near the center, which should have been 5 mG from the power lines, was actually 2 mG due to this passive canceling loop.

When the power line field was reduced by 77% after the utility agreed to reverse-phase the lines, the secondary peak at the rear of the building was no longer detectable.

Shielding from active current loops

If a little is good, more may be better. Why not help the induced counter-current along by giving it some power? This line of thought leads to the installation of a simple sensor/feedback device going to an intentionally placed wire loop. If a magnetic field sensor (such as the MAGCHECK-95) is placed inside a building and hooked up to a feedback amplifier circuit whose output goes to a wire loop strung around the building in a plane that generates a magnetic field that is lined up with the intruding power line field and running opposite to it, a very effective cancellation can be achieved in the central third of the loop area. This is the basis for Ed Leeper's feedback system which does just that. The only problem is that the feedback loop creates a higher-than-ambient magnetic field close to the wire. This could lead to liability in an inhabited area such as a city apartment. Another problem arises due to the field gradient from the source across the loop area. The system works best with net current, whose gradient is more shallow.

This active cancelation effect is routinely used in laboratories to cancel the earth's DC field if that is desired. In this case the current in the loop or coil is DC rather than AC.

Chapter 14

CASE HISTORIES

Paralleled neutrals (GCs)

The following are examples of GCs from different circuits being connected together on the load side of the breaker box in violation of NEC 300-3(b) and 310-4:

• In Los Angeles: a classic case of GCs from two circuits sharing a junction box: the box was at the far end of the house. Each circuit ran on opposite sides of the bedroom, creating a loop effect which produced uniformly elevated magnetic fields. The junction box was found in the attic crawl space and the misconnected GCs were separated according to circuits. The high fields disappeared. A sequel to this job was a call I received from the lighting specialist for this "smart house". He was very happy to find that his weekly calls to deal with problems in his system were no longer necessary. Evidently the magnetic fields from that circuit had been strong enough to induce spurious signals onto his system, which uses the regular wiring to transmit low voltage signals to control the lighting. He purchased a gaussmeter so he could check out other jobs for potential problems.

• In Sausalito, CA: In an office building with at least 40 subpanels and two service entrances we found multiple sources of high magnetic fields. One office suite had particularly high fields. There were several subpanels in the suite, as it had been reconfigured for successive tenants. I found a junction box in the attic heating room where net currents were traced coming and going. See **Figure 14** for a photo of the box with all the GCs wire-nutted together. Not only were some of the GCs from separate branch circuits but some were from different subpanels. This not only produces net currents with their strong magnetic fields along the path of the circuits back to their subpanels but it unbalances the feed circuits for each subpanel. Each circuit now has a net current due to excess or deficit of neutral current and generates a strong magnetic field all the way back to the service entrance, which in this case generated strong magnetic fields in other offices both on that floor and the floor below.

• I was called in by a family living in a desert community north of Barstow, CA who had experienced a sudden onset of electrosensitivity. They were sure something was wrong with the house as they could feel burning sensations, etc. and had rashes on their hands which puzzled their doctor. A child born while they lived at the house was born with three kinds of brain damage.

As this was the first case of electrical hypersensitivity I had come in contact with I was cautious if not a little skeptical. But when I measured the house with the lights on I found elevated fields in the master bedroom, the baby's bedroom, and a field of 82 mG at the couch the family gathered together on to watch TV.

After a couple of hours of tracing I found the point where GCs of two circuits had been connected. As mentioned in the text, it was at a duplex receptacle in the master bedroom where one circuit was controlled by a wall switch and the other was always energized. In such receptacles one must break off the tabs connecting the two parts of the receptacle -- both the hot tab and the neutral tab. In this case only the hot tab had been broken off. As soon as the receptacle was disconnected the field dropped from 82 mG to below a tenth of a mG at the couch. An 82 mG field one foot from the wall is generated by a net current of approximately 12 amps; in this case due mainly to an electric heater used by the family in the winter.

Whether the mother's use of this couch for sleep during pregnancy was related to both the birth defects and hypersensitivity is up to medical specialists to sort out. The hypersensitivity was real as I found by a careful experiment in which I found that the mother was able to tell if a 2 mG field was on or off 11 times out of 12, with no external clues from me or the apparatus I used (a covered coil and a silent mercury switch, plus a predetermined randomized on-off sequence).

Cases like this point out the responsibility electricians may carry if scientific evidence continues to find a connection between certain disease conditions and magnetic fields of over 2 mG or so. Suppose the havoc caused to this family (who have left the house) were actually due to the sloppy and incorrect wiring of that receptacle. And suppose a cause-effect relationship could be shown in a court of law. Electricians need to realize that these kinds of lawsuits are coming.

Neutral/ground shorts

• In an elementary school on Marthas Vineyard, MA: The classic and all-too-common case of a grounded neutral on the load side of the service disconnect (violation of 250.24(A)(5). A fifth grade classroom had a field of 22 mG at desk level running along the middle of the room from an under-floor source. Tracing the path of the field lead to a subpanel where a neutral bus was grounded. Unscrewing the bonding screw in the bus corrected the problem and removed a magnetic field which had existed for years.

• In an old school building in Marin County, CA a major subpanel, which may once have been a service entrance, had a large neutral bus which was not insulated from ground. This was putting a magnetic field of over 4 mG at desk level throughout several classrooms. The electrician replaced the bus with an insulated one, thus removing the field from the classrooms.

• In a school in Napa, CA a subpanel outside a nurse's office had a grounded neutral bus. The neutral current was diverted to a water pipe. A field of 4.5 mG was present at two beds where sick children rest. Typical symptoms of many similar wiring errors in the district's schools were detected and reported. The District Board stated its intention to have the errors corrected.

• In a high school on Marthas Vineyard, MA a neutral/ground short was traced to a receptacle. When the cover was pulled off there was a neat shunting wire going between the neutral and the grounding terminals. This was one of several ways by which a large amount of neutral current was finding its way to the sprinkler pipes which in turn were producing fields of up to 16 mG in the halls and some classrooms. This illegal shunting wire had been intentionally installed by a former electrician as standard practice. When corrections were made by the new electrician the fields were reduced or eliminated.

• A Marin County, CA residence had a strong field running from the breaker box through the kitchen and breakfast nook down the hall to the washer/dryer room. There was also current on water pipes inside the house. I found a bare grounding wire running from the dryer case to a clamp on the copper water pipe. It was carrying the neutral current we were tracing. There were two sources for neutral current on the water pipe in this situation. One was due to the practice of bonding neutral (or GC) to ground in older dryers, with the result that since the optional bare bonding wire was used, the neutral current from the 120V dryer motor was shunted to the water pipe. This case is discussed below.

This brings up an interesting point. NEC 250.60 allows a grandfathered situation in which the neutral conductor of the old style 3-conductor cable of clothes dryers may be used also as the equipment grounding conductor as long as it is insulated from ground all the way back to the service entrance panel. This would cause no magnetic field problems as long as it was done according to the specifications in the Article. However this practice connects the frame of the dryer to the neutral. When an inexperienced installer then adds a supplied grounding wire from the dryer's frame to the water pipe servicing the washer, he lets neutral current shunt to the water pipe. This is a situation the Code article is specifically worded to prevent. Local grounding/bonding wires should not be used on dryers that have internal neutral/ground bonds.

Since the 1996 edition of the NEC eliminated this exception to the general rule of keeping neutral and ground separate on the load side of the main service disconnect (service entrance), it is interesting to ask why this exception was ever made. It had applied not only to clothes dryers but to electric ranges, wall-mounted ovens and counter-mounted cooking units. Note that it had not applied to any unit connected to water pipes. As it has been explained to me, when three-wire dryers were introduced they were connected to the older two-wire circuits which did not have equipment grounding conductors. Since this was during WWII, in order to conserve copper it was not required that the circuits be re-wired with a grounding conductor, and so the neutral was allowed to be used as the ground for the frame as being more secure than running a grounding wire from the frame to a water pipe.

Dryers installed now must be supplied by a four-pronged cable which includes an equipment grounding conductor. Printed instructions on the dryer specify that the neutral-ground bond installed at the factory be disconnected when used with four-conductor cable. These instructions are sometimes poorly worded and so the installer may miss this requirement. Double check.

Neutral to grounding electrode

• Here is an example of the part played by corrosion at the utility's service drop neutral connections: At one time I had an office in a small building on the Oakland waterfront. Soon after I moved in I noticed there was a high magnetic field in the corner area of one of the rooms. The field centered around the breaker box and lead both up and down the wall. I had also been having voltage problems, with one circuit measuring 107V while the other was at 134V. The neighboring office was complaining of computer problems when I operated any machinery.

I went outside and clamped the ammeter around the service drop cable. It read 10 amps net current. I looked for the grounding electrode conductor. Since the building was supplied by plastic water pipes the conductor was clamped only to a ground rod. I clamped the ammeter around the ground rod. It measured 10 amps! How could 10 amps be going to earth through a single 8' ground rod? The answer lay in the fact that being just yards from the Estuary the rod was in contact with the salt water table.

Next I measured the service drop neutral. Only 0.3 amps! So my electrical service was relying on a ground rod and the salt water table to carry the neutral return current back to the transformer. No wonder there were voltage problems.

PG&E agreed to send a linesman out since I mentioned the voltage problem. He came the same day and checked the service drop neutral connection. He found it had been a temporary connection and replaced it with a permanent connector. But no change to the flow of neutral current. Next he went up the pole and wire-brushed and re-clamped the neutral connection there. The problem disappeared. Now there were 10 amps going up the service neutral and only 0.2 amps going in the ground rod. And my voltage was back to 120V on each leg. The neighboring office no longer had computer problems and they became friendly again.

• The next case is unusual in that it involves a child who developed leukemia in the first year of his life. I was called in to measure the magnetic fields in his crib room since it was next to the electric service entrance closet in a large LA estate. I found no unusual field level emanating from the service closet itself, since the equipment was not mounted adjacent to the wall to the child's room, but a substantial magnetic field was coming from an under-floor source which angled out from the service closet location, ran under the crib and exited near the corner of the room. The field at the crib was 4 mG and on the floor where the child crawled it was 12 mG. These values were measured during the day when the only loads were the washer and/or dryer. During the evening the field levels would have been much higher.

The field was due to a net current on the service lateral which ran under the floor. I clamped the ammeter around the copper water supply pipe where it entered the building and found it was carrying amperage. Following my recommendation the owner had a plumber insert a dielectric coupling in the water pipe down near the water meter near the curb, which was about 100' from the house. This took the neutral current off the pipe and restored it to the neutral conductor in the service lateral and balanced the magnetic field, thus reducing the magnetic field in the crib room to tenths.

The child underwent chemotherapy and for awhile was doing well. Before finishing this revision I learned that sadly he had not survived. Was his leukemia due to having been exposed to 4+ mG in his crib and 12+ mG while crawling around on the carpet?

No way to tell about an individual case. However when the major epidemiological studies over the last 10 years or so were examined in a vast review by the World Health Organization's IARC (International Agency for Research on Cancer) in 2001 they came to the conclusion that a doubling of childhood leukemia is seen at an average of 4 mG or more from power frequency magnetic fields. This put it in the official category (2b) of "possible carcinogen".

As an aside to how our governmental agencies are ruled by political forces, the EPA staff came out with the same conclusion more than 10 years ago but the report never made it past the heads of the agency. The draft report was widely read by those in the research community but it never made it out to the public.

Chapter 15

WORKING WITH UTILITY REPRESENTATIVES

The original magnetic field health studies were in relation to power line magnetic fields. Thus public concern focused on power lines. As a consequence many utilities now have EMF representatives who are usually engineers assigned to this duty. Considering that there are over 3000 utilities in the U.S., not all will have a designated EMF person. Company policies vary, but many will send a representative out to a residence if the owner or tenant requests it. They will take some measurements with a gaussmeter and the employee who does this is usually friendly and will take additional measurements requested by the customer. As one utility EMF person explained to me, their job is as much PR as it is to give technical data.

The most common customer concern is with a nearby power line or transformer. From my own experience I have found that only in a minority of cases is the power line producing more than a milligauss at the house and seldom is a transformer close enough to produce a detectable field in the house. Of course when a power line field *is* present it affects most or all of the house. I have measured over 35 mG in an office building at the edge of a transmission line right of way and over 20 mG in houses close to primary distribution lines.

But in many cases what the utility engineer will be measuring in the house are magnetic fields coming either from miswired circuits or neutral currents on water pipes, which involves also a net current on the service drop. The customer naturally asks the utility engineer where these magnetic fields are coming from. The engineer disclaims that they are coming from the power lines but this puts him or her into a dilemma. The customer naturally wants to know what is the source of the fields and the engineer, who is in the residence in the role of an electrical authority may be loath to tell the simple truth, which would be, "I can't tell you specifically, but they have to do with the wiring". After all, utility engineers are not electricians nor are they trained to be experts on building wiring and the National Electrical Code. Their responsibility actually ends at the electric meter outside the house.

But it is hard for someone playing an authority role to say "I don't know", and this leads to complications which the electrician may have to deal with when called in by the customer and told that the utility engineer said this or that.

At a recent EMF conference for utility representatives I was speaking with an engineer whose job was to take magnetic field measurements for his utility's customers who requested it. I asked him what he told the customer when he found substantial fields in the house which were not coming from the power lines. He said he told them that the fields were due to the complex nature of internal wiring, where some circuits might cancel each other's fields and some would add onto each others fields. I stopped him right there and let him know from my experience that if a building is wired correctly the only source of fields in the over 1 mG range, if not external, would be from a water pipe ground situation, a circuit miswiring or from a nearby appliance. The situations he had been ascribing to the inevitable complexity of house wiring were actually due to incorrect wiring. The engineer expressed great interest in seeing this book when it came out, since he has no wish to mislead customers.

So you as an electrician may have to deal with the aftermath of freely given engineers' opinions. Here are a few examples so that you may be forewarned:

• I was called to a house where a designated utility EMF engineer had traced a strong field the length of the house. He was right in concluding that it came from the circuit feeding a subpanel which was distant from the service entrance. His advice to the home owner was to have an electrician re-route the feeder cable above the roof on elevated supports to keep the field away from the living areas!

I found that the subpanel had a neutral bus with a grounding electrode conductor connected to it and leading to a clamp on a water pipe. Evidently this had been the original service entrance, and when the new one was installed at the other end of the house the electrician neglected to remove the grounding electrode conductor from the neutral bus. This bond was in violation of Code and created a dual magnetic field both from the water pipe which ran through the house and from the feeder circuit which now had a substantial net current on it.

The solution was simple and inexpensive: remove the grounding electrode conductor from the neutral bus in the subpanel; a 15 minute job for an electrician. The magnetic field disappeared.

• A second interaction with a utility engineer may be instructive. I had found a situation in a shopping center where two service entrances served a complex of connected shops, including a supermarket. I measured a 20 mG field at a table in a restaurant that served good coffee. It did not take long to realize that the field was coming from a large sprinkler pipe running along the front wall near the ceiling. I traced the pipe to many lateral branches, one of which was heavily loaded and went to the back of a dry cleaning store where the service entrance for the string of shops was located. At the service drop I measured a net current of about 6 amps. At the other service drop supplying the supermarket I measured a similar 6 amps.

The explanation for the current on the sprinkler system seemed clear. The system was common to all buildings. Both service drop neutrals were bonded to the same sprinkler system. Assuming a difference in the impedance in the connections for each of the two neutrals in the service drops, some neutral current from one of the service drops was being shunted through the nice fat sprinkler pipe to the neutral in the other service drop.

Since I had seen this problem in two office buildings, where multiple service drops were bonded to the same water piping system, and since no happy solution had yet been proposed, I decided to call the utility's area EMF engineer and see if he had any suggestions. I also wanted to ask him about the splitting of the bare neutral service drop cable, as I could see that part of it was braided off as a support and connected to the building with an insulated anchor, leaving only a portion of the cable to carry the neutral current. I wondered if the neutral cables were oversized to perform this double function.

When I called I received a surprise. The engineer first denied that this bare conductor could be the service drop neutral, as he believed that service drop neutrals were always insulated! Since everyone knows that the common spun service drops have bare neutrals, I didn't know what to say except have him repeat what he had said so there was no misunderstanding. I told him that the usual service drop neutrals that I see all day long are bare, but then went on to the problem of a water pipe carrying neutral current between service entrances.

His response was that when he finds current on a customer's water pipe, while conducting a magnetic field survey at their request, he recommends that they detach the grounding connection between their electrical service entrance and the water pipe and instead give the water pipe "its own ground rod". I was amazed. There is no clearer violation of Code than what I had just heard. I replied that NEC requires the water pipes in a house to be bonded to the neutral at the service entrance. He countered that that may have been true in the past, but that the 1993 NEC allowed a separate ground for the water pipes. His tone was almost condescending. I held on to my usually volatile sense of outrage and replied that we both better check our books because I had seen no such change in the 1993 Code.

Later on in my office I searched Article 250 in the 1993 Code even though I had closely followed the Code changes as they were reported and discussed. The IAEI Journal is particularly good about reviewing Code changes and I was a member and subscriber. As soon as I assured myself that I was not living in the Twilight Zone I composed a letter to this engineer and informed him of the safety hazards consequent on his recommendation as well as the reason for single point grounding, which is a foundation stone of the NEC. I never received a reply.

By relating these stories I do not mean to say that the average utility engineer would make these errors. The only point I am making is that they have not been trained in wiring practices or asked to pass an exam on the NEC; therefore those who feel the need to tell the home owner what to do about the wiring may get it wrong, in which case you, the electrician, are left to deal first with the misunderstanding and then with the actual electrical problem. Be prepared.

Chapter 16

CITY PROBLEMS

There are problems which arise when buildings are connected to each other, as was mentioned above. Here is another example. I was called into a San Francisco business on the second floor of a 2-story building on fashionable Sutter Street. There were magnetic fields coming from every continuous metallic pathway in the building. There was current on the gas pipes, heating pipes, conduits, air ducts, building columns and girders, as well as on aluminum window frames. I traced some of the paths to the roof, where they crossed over and went to the next building on each side.

The subpanel servicing the business was OK: no net currents on any of the circuits; no grounded neutrals at the box. The large air conditioner on the roof which serviced another business also showed a balanced circuit, even though the duct work was carrying current.

I went to the basement and studied the service entrance equipment which fed two or three businesses in the building. Each of the service entrances had a grounding electrode conductor securely clamped to a copper water pipe. This water pipe was carrying 6 amps. But the 6 amps did not flow on the side of the pipe that led out of the building (the point of supply) but instead flowed into the building. I followed the copper pipe to the wall where the water supply entered the building: it changed to plastic! That section of San Francisco had changed to plastic water mains. All right, this happens, but where was the replacement grounding electrode(s)? There were none. No rod. No alternate grounding electrodes that I could find. However, this did not affect the magnetic field situation.

I turned my attention to the 6 amps on the copper pipe which was headed into the building and also towards the adjacent building. The current split in the maze of pipes: water and gas, as well as conduits. About half the current went through the basement wall into the next building. The rest split up and headed up to the upper floors, and as I had found, crossed the roof in conduits or pipes to the adjacent buildings. Some of it flowed along building steel.

What emerged was a picture of a block of buildings supplied by numerous service entrance cables, each containing a neutral which was connected in damp basements and utility manholes to the power line neutrals. These multiple neutrals would be expected to have varying degrees of impedance; hence neutral current was shunting around between service entrance panels by way of any and all metallic pathways up, down and through the many adjacent buildings on the block, meanwhile creating substantial magnetic fields in the offices. There was a system of heavy iron gas pipes in the building I surveyed, long since disconnected but now carrying amperage throughout the building.

What is the solution to this situation? An individual tenant can do little if anything. I was able to help this tenant by inserting a small piece of plastic between two copper pipes in the basement where they had been touching. Neutral current from the bonded pipe was being transferred by contact to a water pipe supplying the tenant's floor. By separating the two I was able to divert some of the neutral current away from her floor but of course this was increasing it somewhere else. Any real solution would involve coordinated action and cooperation between the utility and the landlords.

As a first step the Code violation should be remedied. Grounding electrodes should be installed to replace the removed (plasticized) water pipe electrodes. But the neutral shunting between buildings would continue. How can this be prevented? As a first step the utility should check and clean all neutral connections serving the service entrances. This may go a long way towards solving the problem. If differences in neutral impedance still exist, and if metallic pipes still present an attractive parallel path for the neutral to return to the transformers, what can be done? Theoretically the piping common to the buildings could be electrically isolated between buildings by inserting dielectric couplers. Bonding at each service entrance would maintain the safety grounding of the pipe segments. However in practice where complex piping systems touch each other and are supported by metallic braces and wires, building steel will be involved, as well as telephone grounds. It is difficult to believe that electrical isolation between adjacent buildings could be achieved and maintained in practice.

The problem of shunted neutral currents between adjacent buildings in cities will have to be addressed by the utilities and other appropriate agencies and solved by cooperative effort when the EMF issue achieves a higher priority. Meanwhile if the utility will check and clean its neutral connections the problem may be minimized.

EPILOGUE

While working with electricians to correct errors which had caused elevated magnetic fields we sometimes discussed the reasons for these errors and their frequency in America's buildings. I have had estimates from experienced electricians that from 60% - 80% of the buildings in this country have errors of the kind that we have uncovered. My experience would tend to support the higher figure. It is rare for me to find a house or commercial building which does not show elevated magnetic fields from at least one miswired circuit. How can we account for the frequency of the errors?

Discussion with electricians yielded the following factors:

• Simple ignorance of the Code or its correct interpretation.
• Lack of understanding of the electrical principles behind NEC Code articles. Electricians are not electrical engineers. And electrical engineers are not usually familiar with wiring practices.
• The desire to save time or materials.
• Convenience.
• Misconceptions about grounding: "the more grounding the better". This can lead to prohibited neutral/ground connections.

One may also wonder: how do these violations get past electrical inspectors? The reasons can be numerous. Many of the violations will not be easily visible during routine inspections. Some inspectors are not electricians and must inspect all building systems; thus they may not be able to do a thorough job with the electrical system. Even experienced inspectors may not realize that a bunch of neutrals wire-nutted together in a junction box may be a violation. Sometimes "that's the way we've always done it" wins out over a close look at the relevant Code article. Local politics may be involved. The inspector's interpretation prevails locally and appeal is made very difficult.

The common use of gaussmeters may change the situation considerably. Once the building has been energized the inspector can return with a gaussmeter and if a circuit has been miswired in the ways outlined in this book, that error will be immediately obvious by the high reading on the gaussmeter. In the future one can expect that a gaussmeter will be a standard tool of electrical inspection. If the electrician has one, he can check his work before the inspector arrives.

It is clear that the profession of electrician is changing fast. The complexities which have been introduced in recent decades have changed the profession from a skill anyone can pick up after working with an electrician for awhile to something approaching a "high tech" profession. I am told by European-trained electricians of the rigorous and lengthy training program they must go through before becoming licensed. Perhaps this country will have to look to raising its standards and upgrading the profession.

Glossary

Note: These definitions are worded for the layman. For more technical terminology for electrical terms I recommend the National Electrical Code Handbook (see Bibliography).

Amperes, amps, A: A unit of measurement for electric current.

Bonding: Connecting solidly. NEC: The permanent joining of metallic parts to form an electrically conductive path that ensures electrical continuity and the capacity to conduct safetly any current likely to be imposed.

Bonding screw: A screw provided by the manufacturer in the insulated neutral busbar of a breaker panel which can be turned in to bond the busbar to the grounded box if the box is used as a service entrance main disconnect. If the panel is used as a subpanel the screw should be turned out or removed.

Branch circuit: A circuit leading from a panel or subpanel to a group of outlets.

Bus or busbar: A bar with screw connections for either neutral or grounding conductors. It can be used for both only if it is a service entrance panel with not more than 6 main switches.

Clamp-on ammeter: A current-measuring instrument whose jaws open to encircle a conductor or cable and measure the amps or milliamps of net current. It is actually a magnetic field detector calibrated to read in Amps.

Code: National Electrical Code (NEC). Revised every 3 years. Published by the National Fire Prevention Association.

Conduits: Metal pipes for carrying electric circuits. They may constitute the equipment grounding conductor for a circuit.

Distribution lines: Lines carrying power to neighborhoods (primary distribution) and to one or several buildings (secondary distribution).

Earth: English term for ground. See Ground.

Electric field: A zone of potential difference between oppositely charged conductors or between conductors and ground. It is measured in Volts per meter or V/m, or kV/m. Electric field is a function of voltage and not of current.

ELF: Extremely low frequency, which is the designation for the frequencies used by the power system. Formerly understood to mean frequencies from 30-300 Hz, but now most often used to mean up to 1 kHz and increasingly up to 3 kHz, which is where VLF begins. The definition depends on the source consulted.

EMF: When standing for electromagnetic fields: several meanings: Now often used to refer to the general phenomena associated with fields spreading out from conductors or antennas, depending on the frequency. At radio frequencies and above, the electric and magnetic fields are coupled together and one can be predicted from the other. At power frequencies the two fields are separate and behave in different ways. One cannot be predicted from the other. Hence "magnetic fields" and "electric fields". Because of the need for a quick label for magnetic field effects, "EMFs" is often used to stand for power frequency magnetic fields. Engineers don't like such usuage, but language changes. Otherwise we would all be speaking like Shakespeare or Chaucer.

Equipment grounding conductor (EGC): The technical name for "safety" or "ground" wires. May also apply to conduits. As conductors they are usually bare or green. They are intended to carry current only when there is an accidental ground fault where a hot conductor is touching grounded metal. Contrary to the usual understanding of the word "ground" the current does not go to actual ground (earth) but continues back to the transformer by way of the service neutral conductor to complete the circuit. A more accurate term would be equipment bonding conductor.

Feed or feeder circuit: A circuit which carries power to a subpanel or a separate building from a service entrance panel.

Gauss: The measure of magnetic field strength (flux density) used in the U.S.A. Equals 1,000 milligauss (mG). In Europe the unit used is the Tesla or microTesla (μT). 1 μT = 10 mG.

Gaussmeter: A magnetic field strength meter which reads directly in mG and/or Gauss. Those designed for industry may not be sensitive enough for environmental measurements. May be analog or digital. See Instrumentation chapter.

GFI: Ground fault interrupter. A device designed for personal protection from shock and fire. It trips when it detects a difference in current between hot and grounded conductors of as little as 5 milliamps. The miswirings documented in this book would have tripped GFIs if they had been installed in those circuits.

Ground: Has so many meanings it must be defined by its context. Used as a verb or noun. As a verb it means to connect in some way to either earth or to a conductor which returns to the service entrance panel and is bonded to the neutral conductor which returns to the transformer. Same as the English use of "earth".
 Sometimes used when the more accurate term would be "bond". Paradoxically, it is not the earth connection which protects a circuit from a "ground fault", but a solid connection back to the transformer neutral. See chapter on grounding.

Ground fault: An unintentional or accidental connection between a hot conductor and a grounded part or conductor. If wiring is correct this will result in a tripped breaker in time to prevent fire. Dangerous shock may occur in the measurable time interval. Only a GFI protects immediately against shock.

GC: see below.

Grounded circuit conductor (grounded conductor or GC): Commonly called the "neutral", even though this term makes sense only when it is used in reference to a multi-phase circuit or when used as the neutral for a 3-wire circuit carrying current from the 2 hot legs. It is the white wire. In a two-wire circuit it carries the same current as the hot conductor, in the opposite direction. In a multi-conductor circuit it will carry only the unbalanced current resultant from the hot conductors.

Grounding electrode: An approved metallic object for making electrical contact with local earth. Commonly used electrodes are water pipes in contact with earth for 10' or more, 8' ground rods driven into the earth, grounded building steel, buried conductor rings ("Uffers"), and certain others specified by NEC Article 250-81.

Grounding electrode conductor (GEC): The heavy conductor which leads from the neutral/ground bonding point at the service entrance to the local grounding electrode. It is usually seen as a heavy solid or braided copper conductor which leads to a clamp on a water pipe or a ground rod, etc.

Hot: As a noun it refers to the ungrounded circuit conductor carrying the voltage. Commonly at 120V to ground, or 240V to its paired hot conductor. Usually colored black, red or blue. Caution: in some situations a white wire may be hot!

Hot leg: One of the two service entrance conductors carrying voltage from a transformer which is fed by a single phase from the primary distribution line. Each hot leg to a residence is usually 120V to ground or 240V to its paired hot leg. Since these two legs are 180° out of phase, they are sometimes called "phases" by electricians and labeled A and B.

Impedance: The quantity (ohms) measuring whatever impedes the flow of electricity. It is composed of resistance and reactance. The reason there is greater impedance in a circuit when paired conductors are separated spacially is that this increases reactance.

Knob-and-tube wiring: A system of routing conductors separately inside a building which was in use until the 1940s and in some cases longer. The hot and neutral conductors are spacially separated by inches or feet and there is no equipment grounding conductor. The separation produces high magnetic field levels. Found in many older buildings which have not been fully upgraded. "Knob" refers to the porcelain support for conductors; the "tube" is the porcelain cyllinder inserted in joists, rafters, etc to allow the conductor to go through.

Linear loads: Electrical devices which are resistive, such as incandescent lights and heating coils. They do not distort the 60 Hz sine wave.

Load: Used to mean either the amperage on a circuit or the appliances, lights, etc. which are using the amperage.

Load side of main disconnect: Anywhere inside the building relative to the main switching device at the service entrance.

Magnetic field: A zone of energy which accompanies every electric current. Also used for the zone around a magnet and the natural DC field of the earth. A DC magnetic field remains steady once turned on. An AC field is in constant flux at a rate depending on the frequency of the system. The strength of the field is a function of current, not voltage. Both AC and DC fields are measured in milligauss or microTesla but only the AC fields are believed to have a significant biological effect at environmental levels. Further research may reveal some DC effects.

MAGSTICK tracer unit: A tool for easily tracing circuits or pipes with net current inside walls and underground. See Instrumentation chapter.

Main disconnect: A technical term for the switch or group of breakers that shuts off power to a building. It may consist of no more than six breakers and is located near the electric meter on the outside or just on the inside of the building.

mG: Milligauss.

Milligauss: One thousandth of a gauss. Used mainly in the USA. In Europe and in many scientific papers microTesla (µT) is preferred. 1 mG = 0.1 µT. Or 1 µT = 10 mG.

Mu metal: Mu stands for "µ" which means magnetic permeability. It is a brand name for a general group of alloys which allow magnetic lines of force to concentrate within them, thus freeing the space outside the shielding of those lines of force which would otherwise constitute a magnetic field. Simply put, this metal is excellent for shielding purposes and is produced in thicknesses down to foil gauge. It has no effect, however, on net current magnetic fields, and so plays no part in reducing the type of fields described in this book from wiring errors. See Appendix A for sources.

NEC: National Electrical Code. Revised every 3 years. Most municipalities and states adhere to it but may add to or modify it for local use.

Net current: The current unbalance left over when two or more conductors running together do not balance each other in amps and/or phase angle. Detected by either a gaussmeter or a clamp-on ammeter clamped around the circuit. The magnetic field produced by net current acts the same way as a field from the same current traveling in a single conductor. The field weakens slowly (1/r).

Neutral, or neutral conductor: The common term for a grounded conductor (GC). Usually white. Some like to reserve the term for denoting the grounded conductor in multi-phase cables, but common usage includes single phase circuits. Also used to refer to the current it carries.

Neutral/ground short: A term used in this book to mean a connection of a neutral or GC to an equipment grounding conductor or other grounded metal where it should not be: on the load side of the main service disconnect.

Non-linear loads: Electrical devices which take bites of power out of the 60 Hz sine wave, changing the wave form and frequency, introducing non-sinusoidal wave forms and harmonics into the entire circuit. Typical sources: computers, dimmer switches, solid state motor controls, fluorescent lights.

Phase: Usually used to refer to one of the voltage-carrying conductors in a power line or in an entrance cable to a building. The word phase means that the AC impulse is timed differently from that of the next conductor, and the difference is expressed as an angle, so that a typical phase angle for 3-phase systems would be 1/3 of 360°, or 120°. This difference in timing in the transmission of a 60 Hz sine wave results in cancellation of the fields between the three phases. Any unbalance is picked up by the accompanying neutral conductor. If some neutral flows elsewhere, such as in water pipes, there will be a net current on the lines.

Radiation: A term for energy which leaves its source and travels in free space. Power frequencies generate only negligible radiated energy due to the low frequency. Instead the fields spread out and collapse back on the source. Some transmission lines are long enough, however, to become better antennas, and so some 60 Hz radiation is detectable from satellites. Radio frequencies and above are true radiating sources.

Reactance: One form of impedance; that is, an impediment to the flow of current. It increases when pairs of conductors are separated, such as when neutral is shunted. Reactance is composed of inductance and capacitance.

Reverse phasing: A simple method of reducing magnetic fields from double-circuit transmission lines. Instead of the two sets of lines having an A, B, C and A, B, C phase sequence, it is changed to A, B, C and C, B, A. Simple but highly effective. A single circuit can be split into two circuits to make use of this cancellation method. This is called **split phasing**.

Service drop: The overhead cable which brings power to your residence from the distribution line. It usually attaches to the eaves of the building. The old service drops were spaced conductors; presently they are "spun" or twisted together and supported by the bare neutral conductor. The old way created higher magnetic fields; the present way is an excellent canceling configuration but does not affect net current fields. If the service is underground the cable is called a **service lateral**.

Service entrance: the equipment, usually on the outside of the building, where the service entrance cable from the service drop or lateral is connected to the meter and main disconnect for the building. Also called **service point**.

Shielding: Methods for reducing the magnetic field strength from some sources in a given area. Active shielding uses an out-of-phase magnetic field to partially cancel the intruding field. Passive shielding uses high-permeability mμ metal or similar alloy to concentrate the field inside the metal. These methods result in measurable field reductions when using a gaussmeter. Net currents are not susceptible to passive shielding.

Short-circuit: An unintentional connection between hot and neutral conductors in a circuit or between two hots from opposite legs or different phases. Should result in the tripping of a breaker. This action is independent of ground connections.

Signal injector: A device which sends an audio frequency signal along a conductor which can be detected by the receiver unit. See Instrumentation chapter.

Single phase service: Service where a single phase from a three-phase primary distribution line goes to a transformer's primary coil. The transformer's secondary coil provides two hot legs from the opposite ends of the coil and a grounded neutral from the center of the coil. These three conductors are twisted together and go to the residence providing 120/240V service.

Supply side: (of the service disconnect): On the side of the main building switch which is on the utility's side of the switch.

System grounded neutral: The neutral of the utility's distribution system which is grounded at the transformer and periodically throughout the distribution system.

Tesla: A unit honoring the man who developed our AC power system as well as countless other EMF uses. The unit is used in Europe and elsewhere to quantify magnetic field strength. Usually used in units of μT (microTesla). 1 μT = 10 mG. Also used in nanoTesla, nT. 100 nT = 1 mG.

Three phase service: Service that brings in all three phases from the power line plus the neutral. Common in commercial buildings.

Three way switch: Refers to lighting circuits controlled by two switches at different locations. May also be 4-way for multiple control points. If incorrectly wired can produce high magnetic fields.

Two phase service: Service that brings in two of the three phases from the power line plus a neutral. Each phase is 120V to ground and 208V phase to phase. This type of service always produces significant neutral current since two phases never balance. Common in commercial buildings supplied with three-phase to the main panel. Two-phase goes out to subpanels.

APPENDIX A

Instrumentation

Note: The **MSI** products described were designed or co-designed by the author but are now produced by a new owner of **MSI** . The author has no financial connection or benefit. Neither ⸍ does he have any financial connection with any of the other sources recommended, other than in the retailing of this book.

1. Gaussmeters

Single axis gaussmeters

If you are going to have only one gaussmeter I recommend a single axis meter with a separate probe. This gives you both the field strength and a directional indication. It must have a resolution of 0.1 mG to be useful. It is also important to have a meter which meets the accuracy of ANSI Standard 644-1987 which specifies 5%. It should be able to measure sources found in buildings, including devices like electric mixers which can be as high as 30,000 mG (30 Gauss).

In addition it should correctly measure not only the fundamental (60 Hz in the U.S.) but also the common third harmonic (180 Hz) since many circuits in offices and in some residences give off high percentages of the third harmonic frequency. A gaussmeter which measures only 60 Hz may miss 50% of the power frequency magnetic field or more, and will completely fail when trying to measure ELF fields from some VDTs, whose frequencies range from 45 Hz to 90 Hz or higher. Gaussmeters which respond to harmonics with a linear response to frequency may indicate a magnetic field 300% higher than what is actually there. Most of the cheaper gaussmeters have this problem.

Devices which produce harmonics are computers, fluorescent lights, dimmer switches, motor controls and many devices which are electronically switched or regulated. These harmonics then characterize the circuits feeding them as well as the neutral currents and the resulting water pipe currents.

One-piece single axis gaussmeters look attractive because they are compact and appear to be easy to use. Ads may say, "Just turn it on and read the field". Not true. The entire meter must be turned randomly until the highest reading is obtained. Only the highest reading is correct. Unfortunately the user has to turn his or her head and body to be able to read the display when the meter is searching for the maximum. For this reason a meter with a separate probe or sensor is essential. The meter is held still while the sensor module is rotated until the maximum reading is obtained. This is the correct milligauss value.

There is a line of misinformation that gets repeated in occasional articles written by non-technical people and which even appears in some government literature intended for consumers. The statement is made that single axis meters can only measure one axis of three axes which exist in nature. This is a misconception. By turning the single axis sensor one lines it up with the natural resultant of all fields present. That is why it reads the same as a 3-axis meter, assuming both are accurate. For a thorough discussion of when a *3-axis* meter may read inaccurately, see the author's article, *The Effective Use of the Single-Axis Meter* in the July/Aug '94 issue of *EMF Health Report*.

I recommend two single axis meters in the under $350 range which meet the above criteria.

The **MSI-95** is highly accurate over a wide frequency range, 30Hz through 500Hz, and a -3dB bandwidth of 15Hz to 4 kHz. The gaussmeter consists of a compact sensor connected by a cord to a digital multimeter. The sensor can also be connected to an oscilloscope and will give an accurate current waveform of the source causing the magnetic field. This can help identify the source, since each type of source has a waveform signature.

The author can vouch for this performance since he co-designed the meter and did all the testing and component specifications. The present owner of MSI was responsible for the basic circuit design. Accuracy is better than ANSI Standard 644-1987. NIST traceable. Prices in the $250 range. Call **(800) 749-9873** for current pricing or go to www.magneticsciences.com

Another good gaussmeter site is LessEMF.com.

For those who prefer an analog (dial face) meter, the **ELF Sense Model 1A** at about $340 is an accurate meter. The sensor is in the larger case with the electronics and must be turned to orient to the field while the small meter face is held still. The range of the meter is 0.1 mG - 1,000 mG (1 Gauss). Accuracy meets ANSI Standard 644-1987. From Expantest Inc.: (207) 871-0224.

3 axis survey gaussmeters

For a fast initial survey the triaxial gaussmeter is very convenient to use. I use one when conducting the first go-around, recording milligauss numbers on a floor plan. A triaxial meter does not have to be oriented, as it calculates the resultant of three sensors inside. The only drawback of a triaxial is that it is more difficult to discover the source of the field compared to using a single axis meter. The range is also necessarily more limited. The **Bell 4080** triaxial gaussmeter sells forabout $260 and has a range from 0.1 mG to 511 mG. The digital display is easy to read and updates about 3 times per second. Its accuracy meets ANSI Standard 644-1987. NIST traceable.
Stocked by www.magneticsciences.com or LessEMF.com.

There is also a popular low cost multi-purpose analog three axis meter on the market which can give very inaccurate results because of its basically linear frequency response and the fact that it does not calculate the resultant in an accurate way. Useful for scouting out fields but you can't trust the numbers.

3 axis recording gaussmeters

If you need a record of the field over time, or need a magnetic field profile of a power line, or if you were to get into 3-D mapping of the field in a large area, there are 3-axis recording gaussmeters available for around $2,000. I am familiar with Enertech's **Emdex II** which records data at rates of 1 ½ seconds or longer and downloads to a PC. The included software program allows printout of graphs and statistical analysis of the data. I use mine to record power line fields over 24 hour periods from one location and also to obtain field profiles over a client's lot when the source is one or more power lines. It is small enough to be worn on the body or carried in a pocket for a personal record of exposure over a time period. The meter is awkward to use for spot measurements because the display is so small and the slow display update rate of 1 ½ seconds introduces abrupt changes compared to the smoothness of the 3 times-per-second of the other meters described. www.enertech.net

Recording with single axis gaussmeters

If you want to know the maximum, minimum and average milligauss value over a period of time up to 36 hours you can plug in the MSI sensor (MAGCHECK) that comes with the gaussmeter to a multimeter which has the recording function, such as the Fluke 83. The sensor plugs into the AC volts jacks with its included double banana plug. The meter must have a 0.1 mV AC resolution to work correctly. The MSI sensor, called **MAGCHECK** can be purchased separately for use with your favorite multimeter. The **MAGCHECK 95** can also be used with multimeters that have a frequency setting and will read out the predominant frequency being measured, as long as it exceeds the minimum resolution of the meter. Check with MSI for meter limitations. www.magneticsciences.com

The **MAGCHECK** sensors can also be plugged into a data logger for magnetic field recording. The sensor puts out 1 mV per mG so the data logger must be able to respond to that voltage level. Extech sells a data logging multimeter for $249: the ML710. Checking the internet I found it on www.testequipmentdepot.com It downloads by RS-232 cable to a PC (95 or 98). For a variety of data loggers try also microdaq.com. Also try Radio Shack for their data logging multimeter.

Wave form display

One of the great things about the **MSI Magcheck 95** sensor is that it can plug into an oscilloscope and give you an accurate current wave form, even with harmonics. It can do this because of the Magcheck 95's flat frequency response across the typical harmonic range.

There are multimeters now which have a display which will show you exactly what you are measuring as well as show the frequency. No need to clip to conductors. Just hold the sensor in the field and see the wave form.

If you are only interested in predominate frequency, it can be plugged into any of the frequency-reading multimeters available today.

2. Net current tracers

MAGSTICK tracer unit

This unit uses a sensor coil and an audio amplifier mounted on a PVC pipe and comes with matched earphones. The amplifier has its own speaker but the earphones introduce great sensitivity and block out other sounds in noisy environments. A volume control tones down the 60 Hz and harmonic hum.

The unit is used to quickly trace circuits and currents on pipes, etc. including underground cables. Wherever there are net currents the unit picks them up loud and clear. The directional coil on the end of the stick can be used to find either maximum or the null point. Though used mainly to vastly speed up the process of tracing faulty circuits, a lot of information comes from actually listening to the sound of the source you are tracing. A fluorescent light circuit source sounds very different from an incandescent circuit. Computers have their own sound. A pure 60 Hz hum is deep and may indicate a transformer or a service cable. The sound will help you separate out different sources.

The MSI unit sells for $240. Some people with time and patience may wish to make a similar device from a Radio Shack telephone listening coil, their hobby amplifier, earphones and PVC plumbing supply parts. Be sure to match the impedance of the earphones to the amplifier.

3. Clamp-on ammeters

To make the measurements I describe in this book you need at least two clamp-on ammeters, and preferably three. They all must have tenths of an amp resolution. Hundredths is even better. (.01 A)

You need a mini one for clamping on individual circuits in panel boxes: I am currently using the Extech 380947. It has a 23 mm opening , measures AC and DC, reads frequency, and for a fluctuating current reads max and min over time. There are others available at meter websites on the internet.

I also like a standard size one like the Fluke 33 true RMS ammeter which can be left on a fluctuating line to record max min and average. I use the average for recording.

Since you may have to clamp around large diameter conductors like water pipes, service entrance cables, and even bussways, it is very useful to have a flexible probe ammeter such as the AEMC 24-3002 or the 30-24-1-100. These have a 24 inch loop and are very accurate. They plg into a diital multimeter. Available at testequipmentdepot.com as well as other websites. One use of these flexible probes is to snake one around a cable in a high-amp service entrance cabinet without getting your hands near the exposed busses. Or snake around the energized bare copper busses. (Safety concerns may make these measurements off limits for you of course).

There are also a number of clamp-on probes which plug into multimeters. Some have wide jaws. Try sites like Extech and others. You don't need the expensive ones designed for high voltage.

4. Signal injectors and tracers

I have tried several injector/tracer combinations and presently use the Tempo 77HP sender with 200EP receiver combo. The receiver can also be used alone to detect when voltage is present. Volume adjustable. The sender has its own battery power. See Tempo Line Tracer at www.specialized.net

Also good is the power trace combo at www.greenlee.com It uses the power of the circuit and the signal can be traced even inside an aluminum or ferrous conduit.

New products are constantly being developed so check the internet or go to electric industry shows for the latest.

The **MSI MAGSTICK** tracing unit can also pick up the signal from the Greenlee signal injector as well as hear the magnetic field from the circuit being traced.

5. Receptacle testers

There are a number of receptacle testers that give info on the circuits way beyond what the simple hardware store tester gives you. The ultimate is the SureTest ST-1THD Circuit Analyzer. It measures so many things that it is best to read all the features yourself. Try www.inspectortools.com, go to electrical tools, then circuit analyzers. Some measurements are: impedance of each conductor, false grounds, voltage drop at 15 and 20 amps, percent of each harmonic up to the 15th! It has been an invaluable instrument for me. It costs about $550.

6. Shielding materials and consulting

Amuneal Manufacturing Corp.; Larry Maltin, Pres. (215) 535-3000. www.amuneal.com
My #1 recommendation.

Eagle Magnetic Co. www.eaglemagnetic.com (317) 297-1030.
Consumer oriented. You can buy sheets or rolls of high permeability shielding foils.

Notes for 2012 edition:

Prices may have changed since this was printed. In addition, there are more internet sites which can be checked out for a proliferation of gaussmeters and for other meters.

LessEMF.com has a large selection as well as a complete selection of books on EMFs and health effects. They also sell earthing procucts (see last note below).

Magneticsciences.com has a reliable selection of gaussmeters and other instruments.

Amazon.com can be searched for a wide variety of products.

FW Bell has produced a new generation of 3-axis gaussmeters.The 4080 and 4090 has been replaced by 4180 and 4190. These are professional grade. One of these plus a MSI-95 single axis gaussmeter gives one professional credebility.

Protective products

Many products which claim to protect one from high magnetic fields strike me as exploitative and phony, though I would have to investigate each one to find out if there is any validity to their claims. My approach is to eliminate the magnetic field. Make sense?

However there has occurred what may be an extremely important health breakthrough, perhaps monumental, in the finding that electrically connecting the body to the earth's free electrons has important and wide ranging positive health effects. See earthinginstitute.net or purchase Earthing through Amazon and other retailers.

APPENDIX B

Bioelectromagnetics 25:102–106 (2004)

Importance of Addressing National Electrical Code® Violations That Result in Unusual Exposure to 60 Hz Magnetic Fields

Jack Adams,[1]* J. Samuel Bitler,[1] and Karl Riley[2]

[1]*Merrimack College, North Andover, Massachusetts*
[2]*ELF Magnetic Surveys, St. James City, Florida*

We evaluated wiring in multifamily developments containing National Electrical Code® (NEC®) violations as a source of unusual exposure to 60 Hz magnetic fields. Two methods were used in this evaluation: measurement and modeling. We measured the building wiring as a source of magnetic fields in six multifamily developments in Michigan. In this small sample, building wiring proved to be an important source of exposure in four of the six cases. In all four cases with exposure from building wiring, one or more NEC violations were involved. To supplement our measurement efforts, we used computer modeling to compare magnetic field exposure due to building wiring with magnetic field exposure from external power lines. Our calculations showed that where the building wiring has a NEC violation leading to net current loops, the exposure due to wiring is likely to be more important than that from external power lines. Our results support the results obtained in a recent study of the exposure of Californian K-12 students to magnetic fields, where building wiring with one or more NEC violation was found to be the single most important exposure source. If 60 Hz magnetic fields are important to avoid, then improved enforcement of the NEC, as required by law, is perhaps the single most important mitigation policy to adopt. Bioelectromagnetics 25:102–106, 2004. © 2004 Wiley-Liss, Inc.

Key words: ELF; exposure source; policy

INTRODUCTION

The question of whether exposure to 60 Hz magnetic fields from sources ranging from appliances to transmission lines has any significant health impact remains a controversial one, even though more than 25 years have elapsed since the first US studies looking at childhood leukemia were carried out [Wertheimer and Leeper, 1979]. Many laboratory studies in the ELF range have reported no significant effects or equivocal effects that are hard to interpret, although some investigators have reported substantial in vitro [e.g., Blackman et al., 2001; Harland and Liburdy, 1997] and in vivo [e.g., Litovitz et al., 1994] effects at intensity levels comparable to typical environmental exposures. Epidemiological studies have also provided mixed results, with two recent evaluations of pooled results for the major studies to date indicating a statistically significant relationship between residential proximity to transmission lines and childhood leukemia [Ahlbom et al., 2000; Greenland et al., 2000]. The backdrop of inconsistent in vivo and epidemiological studies has resulted in a number of suggestions as to what might be appropriate policy, including prudent avoidance [Morgan, 1995] and the precautionary principle [Jamieson and Wartenberg, 2001].

The major ELF research initiative carried out in California over the past 8 years included a study in which the ELF exposure of children in K-12 public schools was assessed [Zaffanella, 2000]. In this study, all school sources excepting appliances were considered; included were transmission and distribution lines, electrical panels, and building wiring. Perhaps, the most surprising result of this exposure study is that building wiring is found to be much more important than all other sources evaluated. Exposure due to the wiring was mostly found to result from National Electrical Code® [National Fire Prot. Assoc., 1999] violations.

The purpose of this study is to look further at the question of the importance of building wiring as an ELF

Grant sponsor: Michigan State Housing Development Authority.

*Correspondence to: Jack Adams, Merrimack College, 315 Turnpike St., North Andover, MA 01845.
E-mail: Jack.Adams@Merrimack.edu

Received for review 22 April 2002; Final revision received 15 May 2003

DOI 10.1002/bem.10155
Published online in Wiley InterScience (www.interscience.wiley.com).

source and at the impact of NEC violations. If building wiring is dramatically more important than external sources, as suggested by the California school study [Zaffanella, 2000], then building wiring needs to be given proper attention relative to other ELF sources. We emphasize that in this study, we conducted measurements on multifamily dwellings, and our results and conclusions apply to multifamily and not single family dwellings.

MATERIALS AND METHODS

There are two parts to the methodology of this study, measurement and computer modeling. We measured the ELF magnetic fields in six Michigan multifamily developments. The computer modeling simulated the ELF exposure from transmission lines and in rooms affected by wiring with NEC violations resulting in net current loops. Before describing the methodology, we provide background information on how NEC violations lead to unusual ELF exposures.

How Violations Can Result in ELF Fields

In typical, correctly wired building circuits the "hot" and the return currents are equal, and the 60 Hz fields from the conductors start out low and drop off rapidly. By contrast, building circuits containing what we term "net current" paths due to NEC violations are the source of ELF fields significantly greater than background. We define "net current," when referring to wiring, as the unbalanced resultant current carried by a circuit when some of its neutral return current has been diverted to another circuit or conducting path. It also applies to the current on the diverted path, which can be metallic paths such as water pipes or another circuit's neutral conductor. In a properly wired circuit, all of the current is on conductors that are in close proximity, and the circuit has no "net current." Net currents are potent because they create a high magnetic field adjacent to the conductors that weakens only slowly, proportional to as one over the distance (d, in meters). The magnetic field B (μT) from a line current source i (A) is B = 0.2i/d. So 5 A net current produces 1.0 μT at 1 m from the circuit.

There are several NEC violations that can lead to unusual ELF fields, and each violation has the same result: some neutral current in a circuit returns by a path other than its dedicated circuit, and thus net currents are present. The most common situation seen in the California school measurement study [Zaffanella, 1999] was where two circuits encircling a large room on opposing sides come together in a common junction box, and the electrician connected the neutrals from the two circuits, resulting in net currents in both circuits. An indepth examination of code goes well beyond the

scope of this study; further details and references are given in [Riley, 1995].

Field Measurements of Multifamily Developments

This portion of the work was done in conjunction with the Michigan State Housing Development Authority (MSHDA). Six developments were selected with an attempt to have a range of construction styles and state of occupancy represented. Five sites were multilevel, usually four or so levels, and one site was townhouse style with separate units. Four structures were not yet occupied, but had energized services so that lights and appliances could be turned on. Two sites were near or within Detroit, and the others were within 100 miles of Detroit. The measurements all took place during July, 2000.

The protocol for a site visit was first to determine if there were significant fields from nearby power lines that would affect the fields within the building. All measurements were made with a Bell 4090 3 axis Gaussmeter. In no case did power line fields have a significant impact. A walk through of the building was then conducted to determine if unusual (>0.05 μT or so) ELF fields were present. In cases where unusual fields were present, the circuit or circuits that had net currents causing the fields were traced down. An attempt was then made to determine precisely what NEC violation caused the net currents by tracing down the source of the net currents in each case.

Computer Modeling

During the course of this work we developed two computer models, one to evaluate the ELF field profile near a transmission line and another to evaluate the ELF fields in a room that has net currents in one or more walls. The software package used to implement the models is MATHCAD, and the transmission and the room models are based on the Biot–Savart law. To evaluate the field profile near a transmission line, the inputs to the MATHCAD document are the location, loading, and phase of all conductors. A simplified model is utilized, where sag is neglected and the lines are assumed straight. Using a similar MATHCAD document, fields in a rectangular room with net currents in up to four walls are calculated. The required inputs are the room length and width, and the location, magnitude, and phase of all net currents.

In this study we looked at field profiles near two hypothetical, single circuit transmission lines, one a 69 kV line with 300 A per phase and the second a 230 kV line with 600 A per phase. A flat, vertical wire configuration was assumed, with 1.83 m between conductors for the 69 kV line and 3.66 m for the

104   Adams et al.

230 kV line. Configurations and loading can vary dramatically. In our case they were based on industry data and chosen so as to yield typical field values adjacent to the lines.

A straightforward hypothetical net room current case was evaluated, where a net current of 5.0 A was assumed in all four walls. The figure of 5.0 A was taken as being a reasonable value, based on net currents observed during the measurement portion of this study and on typical values encountered by one of the authors (KR) doing hundreds of net current measurements across a broad range of structure types between 1990 and 2002. The conductors were assumed to be 2 m above the floor, and exposure was calculated at 1 m above the floor.

RESULTS

Field Measurements of Multifamily Developments

The primary finding in the measurement part of this study is that in four of the six housing developments, significantly elevated EMFs due to NEC violations were measured. In two cases, the industrial sized dryers for building occupants had been incorrectly installed. The manufacturer ships the units with the neutral and equipment wires connected internally to allow for installations in buildings with older 3 prong outlets. However, when installing in a new building, where 4 prong outlets are required by code for industrial dryers, the contractor must remove the internal neutral/grounding connection according to the owner's manual. When the contractor fails to disconnect this internal link, the shorting of neutral and ground creates an alternate path for neutral current, even when the machines are not turned on. We conducted measurements in the hallway near the laundry room and watched the ELF fields incrementally decrease as we sequentially unplugged the machines, which were not in use. Conversely, turning the machines on caused the fields to rise.

In another multifamily development, 3 way switches were wired with 2 wire Romex inside the main utility room, a clear code violation resulting in significant net currents. Fields exceeded 0.4 μT in the unit above the utility room. The building's owner subsequently had those switches rewired, virtually eliminating the fields due to that cause. In a fourth case, net current at most subpanels indicated a code violation, although the precise location was not determined and could possibly have been behind drywall. In this fourth case, the conductors left the subpanels in a bunch rather than individually, which was a code violation for a flush

mounted subpanel and which made diagnosis more difficult.

In two housing developments, NEC violations leading to net currents were not seen. One case was a townhouse style development, within which each unit had its own electric service. Another case was a multifloor development that had no net currents encountered during our visit.

To summarize the results of this portion of our study, in four of six sites unusual ELF fields due to net currents were measured. In the four cases NEC violations were located, and in one of those cases there was evidence of still further violations, which we were not able to find in the time available. Of the two structures that did not have unusual fields, one was the only townhouse development surveyed. Townhouse developments might be expected to avoid some of the net currents seen in the other developments, as there are no large utility rooms, and no large laundry rooms with multiple industrial grade dryers.

Computer Modeling

The calculated 60 Hz magnetic field profiles for two hypothetical, single circuit transmission lines, one at 69 kV and the other at 230 kV, are shown in Figure 1. Note that for the 69 kV line, the calculated fields have dropped to 0.4 μT at 20 m from the line and to 0.2 μT at 30 m from the line. For the 230 kV line, the calculated

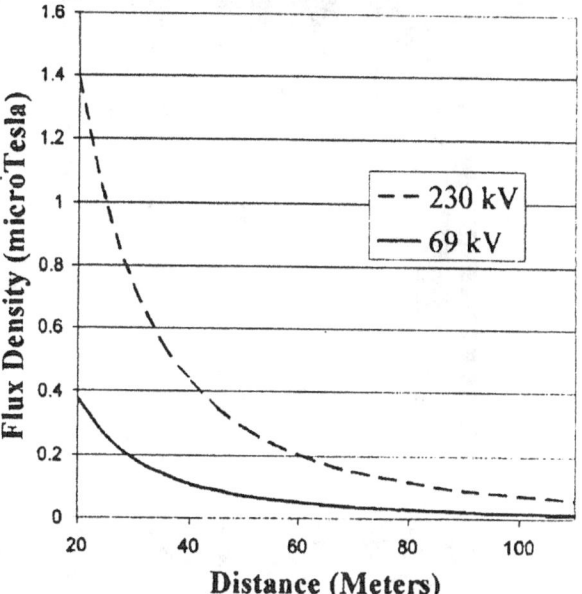

Fig. 1. Field strength in μT vs. distance from transmission line. Upper curve is 230 kV line with 600 A/phase, lower curve is 69 kV line with 300 A/phase assumed.

fields have dropped to 0.4 μT at 50 m from the line, and 0.2 m at 70 m from the line.

The calculated fields for a room with 5.0 A net currents in all four walls are shown as a contour plot in Figure 2. Note that the calculated fields are above 0.3 μT throughout the room and exceed 0.75 μT in about half of the room.

DISCUSSION AND CONCLUSIONS

The measurement portion of our study, while it is it is a very small sample, is along the lines of the California study result [Zaffarella, 2000] that net currents are an important exposure source in 20% of classrooms. In addition, we did not see any examples of ELF fields from power lines impacting exposure inside the structures visited, which also is along the lines of the results seen in the California study.

The ELF field calculation part of the study gives insight into why net currents show up as an exposure source more often than power lines. Fields directly under power lines can be fairly high, but drop off roughly as $1/R^2$, while in a room enclosed by net currents the fields drop off as $1/R$ or even more slowly, resulting in higher exposure where people are. When "typical" power lines are compared to "moderate" net currents, the net currents appear to be significantly more important to address from an ELF exposure stand-

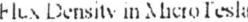

Flux Density in Micro Tesla

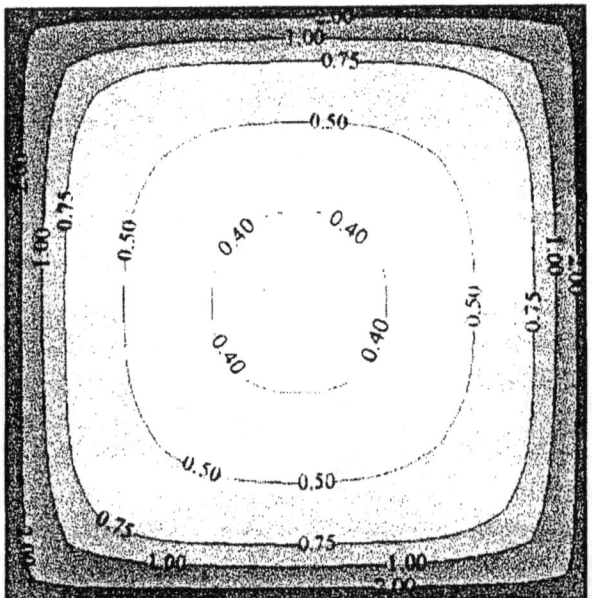

Fig. 2. Contour plot of room 15 m², with net current of 5.0 A in all four walls. [Color figure can be viewed in the online issue, which is available at www.interscience.wiley.com]

point. What we did that was not part of the CA school measurement study was to model the variation of fields throughout the room.

Because in the present study, multifamily dwellings in Michigan have been measured, while the California study dealt with schools, our work indicates that NEC violations leading to net currents may be common across many building types throughout the United States. We note that our results do not necessarily apply to single family residences, where net currents in nearby distribution lines and grounding currents have been found to be important [Zaffanella, 1993].

POLICY IMPLICATIONS

We believe that there are two policy implications stemming from the results of this work and the California School Measurement Study [Zaffanella, 2000]. First, if a comprehensive policy addressing possible health effects of ELF magnetic fields is being considered or adopted, fields due to building wiring need to be included as an exposure source. Second, the presence of ELF fields, at least in larger buildings, can often indicate that NEC violations need to be addressed for reasons of shock and fire hazard. This is actually not a coincidence, since the code seeks to avoid the much higher than necessary inductance associated with net currents. This increased inductance is accompanied by high ELF fields. It also can lead to inductive heating and can compromise circuit breaker response.

If a policymaker chooses to address the potential health effects of long term exposure to ELF magnetic fields, then it appears that net currents in building wiring are at least as important a source to address as external power lines. First, it appears likely that exposure due to building wiring is much more common than exposure due to power lines. Second, a policy to deal with new construction is quite inexpensive and straightforward, requiring a simple additional test for electrical inspectors and education of electrical contractors; in cases where errors are discovered, the fixes are generally quite inexpensive. By contrast, policies to deal with ELF exposure from existing or even new power lines using engineering approaches are not straightforward, involving numerous factors such as visual impact, property values, and effects on the reliability of a system.

In contrast to almost all other ELF sources, a policymaker who places no weight at all on potential health effects due to ELF fields may still choose to address NEC violations leading to net currents. From the perspective of restricting concern to adherence to the NEC, required by both local and federal law in many cases, unusual building wiring ELF fields typically indicate NEC violations resulting in net currents. The net cost of a policy to address this particular type of error may well be less than zero: the upfront costs are quite low as mentioned above; and in exchange for these modest costs, which would become less over time as electrical contractors learn to avoid some simple NEC violations, potential fire and shock hazards can be avoided.

A straightforward and in our assessment prudent policy is to include a check for NEC violations that lead to net currents as part of electrical inspections. The policy advantage of avoiding NEC violations is that shock and fire hazards are mitigated, and potential health advantages are a bonus, rather than controversial. As part of any comprehensive policy seeking to deal with the potential hazards of EMFs, building wiring and NEC violations need to be included.

To summarize this study, our work supports the hypothesis that building wiring, excluding single family residences, is at least as important as and probably significantly more important an ELF field exposure source as are power lines. Both our limited measurement data and our computer modeling data support this hypothesis. From a policy perspective, NEC violations in buildings represent a much more straightforward situation to address than the presence of transmission and distribution lines, where a wide range of complicating factors come into play. Finally, the addressing of NEC violations represents one of the few cases where adopting engineering based ELF mitigation strategy and requiring adherence to local and federal law are directly in alignment.

ACKNOWLEDGMENTS

The authors thank Indira Nair of Carnegie Mellon University, Robert Goldberg of Information Ventures for valuable discussions. We thank the reviewers for valuable comments. We thank Bruce Jeffries of the Michigan State Housing Authority for technical support. We thank the Michigan State Housing Authority for funding to carry out this work.

REFERENCES

Ahlbom A, Day N, Feychting M, Roman E, Skinner J, Dockerty J, Linet M, McBride M, Michaelis J, Olsen JH, Tynes T, Verkasalo PK. 2000. A pooled analysis of magnetic fields and childhood leukaemia. Br J Cancer 83(5):692–698.

Blackman CF, Benane SG, House DE. 2001. The influence of 1.2 μT, 60 Hz magnetic fields on melatonin- and tamoxifen-induced inhibition of MCF-7 cell growth. Bioelectromagnetics 22: 122–128.

Greenland S, Sheppard AR, Kaune WT, Poole C, Kelsh MA. 2000. A pooled analysis of magnetic fields, wire codes, and childhood leukemia. Epidemiology 11(6):624–634.

Harland JD, Liburdy RP. 1997. Environmental magnetic fields inhibit the antiproliferation action of tamoxifen and melatonin in a human breast cancer cell line. Bioelectromagnetics 18:555–562.

Jamieson D, Wartenberg D. 2001. The precautionary principle and electric and magnetic fields. Am J Public Health 91(9):1355–1358.

Litovitz TA, Montrose CJ, Doinov P, Brown KM, Barber M. 1994. Superimposing spatially coherent electromagnetic noise inhibits fields-induced abnormalities in developing chick embryos. Bioelectromagnetics 15:105–113.

Morgan MG. 1995. Fields from electric power: What are they? What do we know about possible health risks? What can be done? Pittsburgh, PA: Carnegie Mellon University Dept. of Eng. and Pub. Policy.

National Fire Protection Association. 1999. National Electrical Code®, First Edition. Clifton Park, NY: Delmar Learning.

Riley K. 1995. Tracing EMFs in Building Wiring and Grounding. Tucson, AZ: Magnetic Sciences International.

Wertheimer N, Leeper E. 1979. Electrical wiring configurations and childhood cancer. Am J Epidemiol 109(3):273–284.

Zaffanella L. 1993. Survey of Residential Magnetic Field Sources, Volume 1: Goals, Results and Conclusions. Palo Alto, CA: Electric Power Research Institute. Report TR-102759-V1.

Zaffanella L. 2000. Electric and Magnetic Field Exposure Assessment of Powerline and Non-Powerline Sources for California Public School Environments, Final Report. Vols. 1 and 11. Berkeley, CA: California Department of Health Services Publications.

BIBLIOGRAPHY

Books

Health effects, exposure data and the controversy

Note: *Most of the books listed have detailed bibliographies of the health research. For that reason this bibliography does not contain references to the primary research, which includes thousands of studies.*

Becker, Robert O. *Cross Currents* (J. Tarcher, 1990). By one of the most experienced researchers in this field; well known for his success in using minute electric currents to heal bone fractures. A thorough review of the evidence for biological effects of power-frequency magnetic fields together with recommendations for maximum permissible exposure. A preceding book, The Body Electric, contains a fascinating professional biography as well as much biological information on EMF effects.

Brodeur, Paul. *Currents of Death* (Simon & Schuster, 1989). Most of this material originally appeared in three articles in the New Yorker magazine in June, 1989.
.................*The Great Power-Line Cover-Up* (Little, Brown, 1993).

Evans, John. *Mind, Body and Electromagnetism* (Element Books Ltd, England, 1986) ISBN 0 906540 86 0.

Health Effects Institute (HEI) *Do Electric or Magnetic Fields Cause Adverse Health Effects? HEI's Research Plan to Narrow the Uncertainties* (HEI, June, 1993. 141 Portland St., Suite 7300, Cambridge, MA 02139). This 130-page document contains a good theoretical introduction to electric and magnetic fields along with a review of the health effects research. The authors are aware of magnetic fields from water pipes though typically have not yet learned about the contribution of wiring errors to magnetic fields in buildings.

Levitt, Blake B. *Electromagnetic Fields ....A Guide to the Issues and How We Can Protect Ourselves* . A thorough discussion of the issues, technical, medical and social. (Harcourt Brace, Spring, 1995).

Pinsky, Mark A. *The EMF Book. What You Should Know About Electromagnetic Fields, Electromagnetic Radiation, and Your Health* (Warner Books, 1995).

Smith, C. W. S. & Best, S. *Electromagnetic Man* (J.M. Dent & Sons, London, 1989). Brings together a great deal of research on the electrical nature of physiological action. A thorough reporting on electric allergies, their diagnosis and treatment.

Sugarman, Ellen. *WARNING: The Electricity Around You May be Hazardous to Your Health: How to Protect Yourself from Electromagnetic Fields,* Third edition, 2004. Originally published by Simon and Shuster. Updated and revised and re-published by Miriam Press. The first book written for the consumer with the emphasis on what you can do to reduce your exposure. With this revision it is the most up-to-date and readable of the consumer-oriented books. Available at www.emfwarning.com or call 315 446-1944.

Technical

The IAEI Soares Book on Grounding, 4th Ed. The International Association of Electrical Inspectors (IAEI). The accepted "Bible" on grounding. The 3rd Edition is my favorite as it seems to contain more complete information.

Holt, Mike. The best source for books on all aspects of electricity, wiring, grounding and bonding can be found at www.mikeholt.com . Mike writes for the electrical Trade magazines and gives seminars around the country. He has been a leader in educating electricians, and carries on the constant fight to help them distinguish between the functions of grounding (to the earth) and bonding (completing the circuit to the source transformer).

National Electric Code, 2005. (National Fire Protection Association, Quincy, MA.)

Seevers, O. C. *Ground Currents and the Myth of Stray Voltage* (The Fairmont Press, Prentice Hall, 1989).
.... *Unique Power System Problems -- Solved* (Fairmont Press, Prentice Hall).

White, D.; Barge, J. M.; George, E.; Riley, K.*The EMF Controversy & Reducing Exposure from Magnetic Fields* (Interference Control Technologies, Inc.,1993, Rte. 3, Box 2000D, Gainesville, VA 22065). A thorough pragmatic assessment of the methods available for reducing magnetic fields from power lines as well as internal wiring and appliances, with legal implications for industry.

Papers

Maurer, Stewart. *Ground Current Magnetic Fields* (New York Institute of Technology. Paper presented at the EPRI Science and Communication Seminar, San Francisco, Oct., 1992).

Videos

Tracing Magnetic Fields in Building Wiring by Karl Riley. 23″ Graphically illustrates the causes of net current fields from wiring errors and shows how to trouble shoot and fix the circuit connections. Available from the author at kriley3@earthlink.net Also from www.LessEMF.com , www.magneticsciences.com , and www.mikeholt.com .

Websites

For unbiased news about EMFs in all frequencies try www.microwavenews.com
To purchase this book: www.lessemf.com , www.magneticsciences.com , www.mikeholt.com

About the author

Karl Riley is the designer of the MSI series of gaussmeters as well as an electric field sensor. He has been an EMF consultant since 1989; his company is ELF Magnetic Surveys. Mr. Riley has pioneered in developing methods of measurement, diagnosis and mitigation of magnetic fields generated inside buildings. He is the first to bring to the attention of the EMF community the central role that wiring errors play in the generation of magnetic fields in buildings.

He is the author of the MSI *Magnetic Field Guide* and author of the chapter on Building Wiring and Grounding in *The EMF Controversy and Reducing Exposure from Magnetic Fields,* Interference Control Technologies, 1993. He has written articles on EMF for U.S. Tech, Indoor Air Review, Compliance Engineering, EMF-EMI Control, and EMF Health Report. Co-Author of *Importance of Addressing National Electrical Code Violations that result in Unusual Exposure to 60Hz Magnetic Fields,* Bioelectromagnetics 25:102-106 (2004). Reprinted here in Appendix B.

He has lectured on EMF measurement and mitigation: at a 1993 DOE-EPA sponsored meeting on EMF in schools, at the June 1993 ICT conference on EMF, at the 1993 meeting of the EMR

Alliance, and at several environmental workshops for professional home inspectors. He has conducted a series of EMF mitigation workshops for electricians, engineers and EMF consultants.

Mr. Riley was featured in a Good Morning America 3-part series on EMF in relation to his work at a Marin County school district in California. He has appeared on several Bay Area TV news segments on EMF.

He currently serves as a moderator on the Code Forum on the www.mikeholt.com website. He currently lives in North Carolina and Marthas Vineyard, MA. He continues to respond to calls for consultation on EMF problems throughout the country. Email: kriley3@earthlink.net .

END

www.ingramcontent.com/pod-product-compliance
Lightning Source LLC
Chambersburg PA
CBHW081549170526
45166CB00009B/2638